and inspectors have attended his courses throughout the US and abroad. His field troubleshooting skills have been valuable in determining where problems occur and where training should be focused to avoid them. He is actively involved in the developments of PV codes and standards in the US and internationally.

PV and the NEC

PV and the NEC presents a straightforward explanation of the National Electrical Code (NEC) in everyday language.

Used throughout the United States and many other countries, the NEC is the world's most detailed set of electrical codes pertaining to photovoltaic (PV) systems. This new edition is based on the 2023 NEC, with most of the interpretations and material staying true long after. It interprets the distinct differences between previous versions of the NEC and the 202 NEC and clarifies how these code changes relate specifically t PV installations. It includes Energy Storage Systems (ESS) a EV to Grid (EV2G).

Written by two of the leading authorities and educators i field, this book will be a vital resource for solar profession well as anyone preparing for a solar certification exam.

Sean White teaches customized solar PV courses for N credit globally. He is a NABCEP Associate Provid NABCEP Registered Continuing Education provider authored nine solar PV and energy storage books a working on more. He teaches courses for various e teaches NEC workshops at SPI and Intersolar. Sea IREC Trainer of the year and the SNEC Online Decade in 2020.

Bill Brooks PE, has 35 years of experience design evaluating grid-connected PV systems. More th

PV and the NEC

3rd Edition

Sean White and Bill Brooks

Routledge
Taylor & Francis Group
LONDON AND NEW YORK

earthscan
from Routledge

Designed cover image: Sean White

Third edition published 2023
by Routledge
4 Park Square, Milton Park, Abingdon, Oxon, OX14 4RN

and by Routledge
605 Third Avenue, New York, NY 10158

Routledge is an imprint of the Taylor & Francis Group, an informa business

First edition published by Routledge 2018
Second edition published by Routledge 2020

British Library Cataloguing-in-Publication Data
A catalogue record for this book is available from the British Library

ISBN: 978-1-032-01918-5 (hbk)
ISBN: 978-1-003-18099-9 (pbk)
ISBN: 978-1-003-18986-2 (ebk)

DOI: 10.4324/9781003189862

Typeset in Times New Roman
by Newgen Publishing UK

Contents

Introduction

Inverter noise on an oscilloscope is on the cover of the 2023 NEC!

At least we think that may be the case, or could it be what a lithium-ion sees when crossing over. We are sure that it must be renewable energy related since we have confidence in our significance.

Photovoltaic (PV) and energy storage systems (ESS), including the types with wheels, are growing fast. Similarly, the PV and ESS material in the National Electrical Code (NEC) is changing faster than anything the NEC has seen since the days of Thomas Edison and Nikola Tesla hashing it out over dc vs. ac. It appeared that Tesla was right when 2-phase ac power[1] was installed at Niagara Falls: ac was the way of the future, but the future is always unpredictable, and with PV and energy storage, dc is making a comeback.

This book is designed to relay to the layperson working in the PV industry the NEC PV-related material and changes as simply as possible. We hope that professional engineers (PEs) and sunburnt solar installers alike will be able to comprehend its writing style and be entertained just enough to not be bored learning about a Code that has been known to work better than melatonin on a redeye flight.

We consider ESS to be under the domain of this book, since Article 706 ESS was birthed from Article 690 PV Systems in the 2017 edition of the NEC.

Since this book is mostly about PV, rather than starting at the beginning of the NEC, we will start with the most relevant article of the NEC, which is **Article 690 Photovoltaic (PV) Systems**; we will then cover the new **Article 691 Large-Scale Photovoltaic (PV) Electric Supply Stations**, which modifies Article 690 for large PV systems, and then dive into the interconnections of **Article 705**

DOI: 10.4324/9781003189862-1

Figure 0.1 1895 Niagara Falls power plant.
Source: Courtesy Wikimedia, https://en.wikipedia.org/wiki/Adams_Power_Plant_Transformer_House#/media/File:Westinghouse_Generators_at_Niagara_Falls.jpg

Interconnected Electric Power Production Sources, which shows how PV and other power sources can connect to and feed other power sources such as the utility grid. The next articles we will cover are those on energy storage: the newer and relevant **Article 706 Energy Storage Systems**; and the older and less relevant **Article 480 Stationary Standby Batteries**. While we are on the subject of energy storage, we will cover **Article 625 EV Power Transfer Stations** (charging and discharging of EVs) and **Article 710 Stand-Alone Systems** (which was formerly 690.10). Articles 706 and 710 were first brought into the NEC in 2017. We will then go back to the beginning of the NEC and look at Chapters 1 through 4, which apply to all wiring systems, including PV. We will see that, in covering the new and renewable PV-centric articles, we already covered the more important parts of Chapters 1 through 4 used for PV systems and all electric installations, such as Article 250 Grounding and Bonding, and Article 310 for wire sizing. There will be many times when we are covering material in Article 690 where we will go back and forth to other articles, since this is the way to properly use the NEC. We will also cover relevant parts of

Chapter 9 Tables and a bit about how to use the Informational Annexes.

Interestingly, all the ".2" definitions in the NEC, such as 690.2, 705.2 and 706.2 definitions, have been moved to Article 100 where all definitions in the Code are now located. We have some of the more important and difficult to explain definitions at the beginning of Chapter 11 on page 263. We also included some industry-relevant abbreviations and acronyms which you may refer to in your journey throughout this book. This book is 51% more likely to be skipped around in rather than read through beginning to end.

The NEC is updated every three years with a new Code cycle. This edition of *PV and the NEC* reflects the 2023 NEC and will discuss earlier versions of the NEC. When the 2026 NEC comes out, this material will not be obsolete; in fact, around half the PV in the United States is installed in places that adopt the NEC three years after a Code is released. For instance, the state with most of the solar in the US is California, and in California, the 2020 NEC was adopted in 2023 and will be used until the 2023 NEC is adopted in 2026. It is also interesting to note that the proposals for changes to the NEC are crafted three years earlier, so the material in the 2020 NEC was proposed in 2017 and will be used on a regular basis by installers and inspectors up to nine years later. Since the equipment changes so fast in the PV industry, the Code writers intentionally leave parts of the Code open-ended to make way for new inventions that you may come up with, which will save lives and may make you rich. We even have places like NYC and Indiana where the NEC is adopted over a decade after it comes out. To see a variety of different sources for when and where the NEC is adopted, you need look no further than Sean's website at www.solarsean.com/nec-adoption-lists-and-maps.

The 2023 NEC proposals for Article 690 and for other solar-relevant parts of the Code were first proposed at meetings in 2020 and put in a Word document by Bill Brooks. This Word document grew, and the proposals were refined with a lot of input. These future Codes were later proposed to the top dogs at the National Fire Protection Association by Ward Bower and Bill Brooks of NEC Code Making Panel 4 in Hilton Head, North Carolina. Ward Bower, our hero and nice guy, invented the grid-tied inverter in 1977 at Sandia Labs and is an endorser on the back cover of this book.

Now is the time to take out your 2023 NEC and follow along to understand PV and the NEC. We have heard of people buying this book instead of the NEC and having inspectors pass them based on the interpretations herein. See if you can get your state to adopt *PV and the NEC*, rather than the NEC.

Here are some different places where you can find the NEC:

1. National Fire Protection Association (NFPA) website free and paid electronic versions, which can only be accessed when you are online. Hard copy versions are also available on the NFPA website. We were able to pay for a PDF version of the NEC until the 2020 NEC was published, at which time NFPA stopped offering the PDF, which was unfortunate for those of us who like to work offline.
   - www.NFPA.org/70
2. Online and brick bookstores such as Amazon or any builder's bookstore.
3. Online places where NEC access is granted after it is adopted. Once the NEC is adopted, it becomes law and it is our understanding that once something is law it can be shared freely.
   - www.up.codes
4. Some building departments, such as the one in the authority having jurisdiction (AHJ) of Tool, Texas, which you can find with a search.
   - www.archive.org

We are sure that NFPA does not like you getting the information in other places, and we are recommending using the NFPA website to get your copy, so we can stay on their good side.

If you are buying a copy which you want to use for a NABCEP exam, you should find out which version NABCEP will be using at the time of your exam and be sure to purchase the NEC Codebook, and do not use tabs or any handwriting except for a highlighter. The NEC Handbook has more details, explanations, and pretty-colored drawings, but is not allowed on NABCEP exams. If you are taking an electrician's exam, check with your state licensing board to see the requirements, and good luck on passing!

We strive for perfection, but just in case, we have textbook corrections at www.solarsean.com/textbook-corrections.

This book is dedicated to your local building inspector.

Thanks for joining Bill and Sean's excellent adventure. Read on and be excellent to each other!

Note

1 The first power plant at Niagara Falls had two phases that were 90° out of phase with each other (weird) (see https://en.wikipedia.org/wiki/Two-phase_electric_power). Now we use single (split), or three phases that are 120° out of phase with each other. This is interesting!

1 Article 690 Photovoltaic (PV) Systems Overview and 690 Part I General

Article 690 first came out in a little book known as the 1984 NEC and has been updated and mostly lengthened ever since. However, with the 2014 and 2017 versions, new articles were made with material from 690, which shortened 690. Taken out of 690 were 705 Interconnected Power Production Sources in the 2014 NEC, and then in the 2020 NEC we separated out 706 Energy Storage Systems, 710 Stand-Alone Systems, and 712 DC Microgrids. In the 2023 NEC, DC microgrids was combined with AC microgrids in 705.

In comparing the original 1984 version of Article 690 to today's NEC, there are many similarities, yet also quite a few differences. Time to dig in!

Let us first list what we are dealing with in Article 690 before we dig deep. This will give us perspective and familiarize us with how to look things up quickly.

The NEC, also known as NFPA 70, is divided into Chapters, then Articles, and then Parts and Sections.

For example, rapid shutdown requirements are found in:

NEC **Chapter** 6 Special Equipment
Article 690 Solar Photovoltaic (PV) Systems
Part II Circuit Requirements
Section 690.12 Rapid Shutdown of PV Systems on Buildings

DOI: 10.4324/9781003189862-2

Figure 1.1 1984 NEC (a much smaller Codebook).
Source: Sean White.

Here is what we find in Article 690:

Article 690 Solar Photovoltaic (PV) Systems
Part I General
- 690.1 Scope [Section 690.1.]
- 690.2 Definitions no longer exist. [2023 NEC moved all ".2" definitions to Article 100 Definitions.]
- 690.4 General Requirements [They could not come up with a better title for this "grab bag" category.]
- 690.6 Alternating Current (ac) Modules

Part II Circuit Requirements
- 690.7 Maximum Voltage
- 690.8 Circuit Sizing and Current
- 690.9 Overcurrent Protection [Article 240 is also Overcurrent Protection.]

690.11 Arc-Fault Circuit Protection (Direct Current)
690.12 Rapid Shutdown of PV Systems on Buildings [Bill coined this term.]
Part III Disconnecting Means
690.13 Photovoltaic System Disconnecting Means
690.15 Disconnecting Means for Isolating Photovoltaic Equipment
Part IV Wiring Methods and Materials
690.31 Wiring Methods
690.32 Component Interconnections
690.33 Mating Connectors
690.34 Access to Boxes
Part V Grounding and Bonding [Article 250 is also Grounding and Bonding.]
690.41 PV System Dc Circuit Grounding and Protection
690.42 Point of PV System Dc Circuit Grounding Connection
690.43 Equipment Grounding and Bonding
690.45 Size of Equipment Grounding Conductors
690.47 Grounding Electrode System [Experts argue over a lot of this article, which is interesting to observe.]
690.50 Equipment Bonding Jumpers [Removed in 2023.]
Part VI Source Connections [Previously Part IV was "Marking," and Part VI Marking was removed, and the former Part VII was promoted to Part VI. Just because Part VI marking was removed does not mean the requirements were removed, they are mostly accounted for in other places as we will explain later in this chapter.]
690.51 Modules and Ac Modules [Removed in 2023.]
690.53 Dc PV Circuits [Removed in 2023.]
690.54 Interactive System Point of Interconnection [Removed in 2023.]
690.55 Photovoltaic Systems Connected to Energy Storage Systems [Removed in 2023.]
690.56 Identification of Power Sources [Directs us to 705.10. Rapid Shutdown Sign moved to 690.12.]
690.59 Connection to Other Sources [Directs us to Article 705.]
Part VIII Energy Storage Systems [previous Part VIII removed, and material below put in Part VI Connections to Other Sources in 2023 NEC.]
690.71 General [Removed.]
690.72 Self-Regulated PV Charge Control

Now it is time to dive into the detail of Article 690.

Article 690 Solar Photovoltaic (PV) Systems Part I General (part)

690.1 Scope (section 690.1)

Word-for-word NEC:

> 690.1 Scope. This article applies to solar PV systems, other than those covered by Article 691, including the array circuit(s), inverter(s), and controller(s) for such systems. The systems covered by this article include those interactive with other electric power production sources or stand-alone, or both. These PV systems may have ac or dc output for utilization.

Big image changes! The images at the beginning of Article 690 that we have seen over and over for a long time have changed. The images with ac and dc coupled systems were moved to the beginning of Article 705, with different terminology explained later in this book. Then the new images, which were drawn by the artist known as Bill Brooks in Informational Note No. 1 / Informational Note Figure 690.1, cover some very noticeable terminology changes as we see here in Figure 1.2 of this book on the following page.

2023 NEC 690.1 Informational Note No. 2 is self-explanatory and was called Informational Note No. 1 in the 2020 NEC: "Article 691 covers the installation of large-scale PV electric supply stations."

Article 691 Large-Scale PV Electric Supply Stations was introduced in the 2020 NEC and covers PV projects over 5MW output.

Discussion: Before the 2023 NEC we called circuits directly connected to solar cells (wild PV) "PV source circuits" before they were connected in parallel in a dc combiner (a.k.a. combiner box), and "PV output circuits" after they exited the dc combiner. The 2023 NEC changes this terminology and calls the circuits entering the dc combiner either a PV source circuit or a PV string circuit.

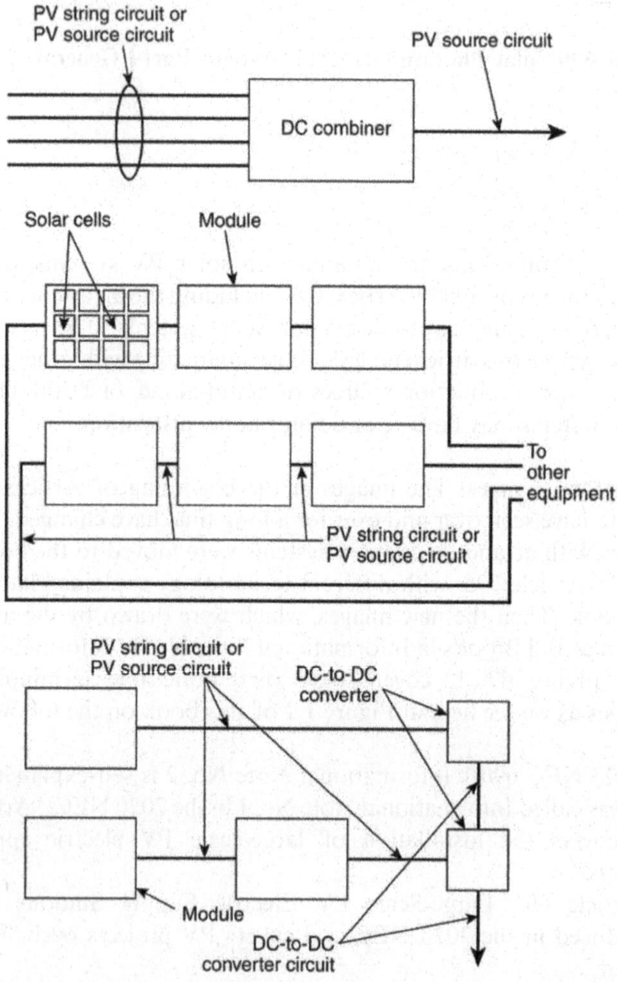

Figure 1.2 2023 NEC Informational Note Figure 690.1 by Bill Brooks, NEC Code Making Panel 4.

Summary of PV circuit names

Before entering a dc combiner:

2023 NEC: PV string circuit or PV source circuit

2020 NEC and earlier: PV source circuit

After exiting a dc combiner:
2023 NEC: PV source circuit
2020 NEC and earlier: PV output circuit

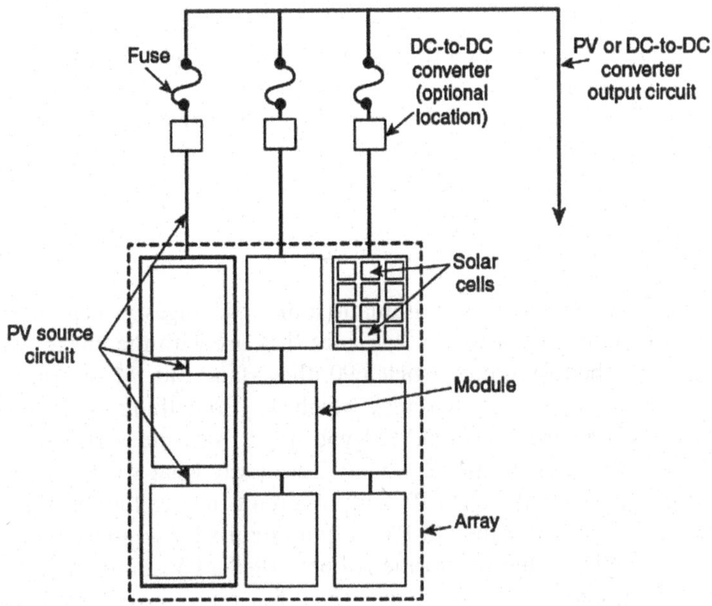

Figure 1.3 2020 NEC Informational Note Figure 690.1(a) identification of PV power source components for comparison of changes.

Courtesy NFPA.

String Theory

We can now call a PV source circuit a "string." The term string has been in the IEC (International Electrotechnical Commission) and many other international codes and standards since the beginning of solar time. Since everyone in the industry is calling PV source circuits "strings," would it be acceptable to call a dc-to-dc converter source circuit a "string"? Perhaps, if you are into stringing together slang.

On another note, we often hear installers calling microinverters that are connected in parallel a "string." We need to correct the microinverter "string" concept by calling it a "branch" rather than a "string," since microinverters connected together are generally connected in parallel, like a branch circuit, and do not have that "series-string thing" happening.

There is microinverter technology that consists of lower-voltage microinverters connected in series with other microinverters so that the voltages add, but it is not currently on the market.

There are some rather dramatic looking changes, which really do not change what we do, but since they are with the images we first see when we go to Article 690, they stand out. Most noticeably, we see PV string circuits, which was something every diligent solar instructor would tell you not to say and was a slang word, just like solar panel is now. Essentially what we are doing is getting rid of the term PV output circuit then calling both PV source circuits and the old PV output circuits PV source circuits. We are calling what was formerly only called a PV source circuit a PV string circuit. We will go over this a few ways, so you can put it in your long-term memory.

Wild PV!

When solar cells are in a circuit with no electronics, such as dc-to-dc converters or inverters taming the PV, we can call it "wild PV" according to Wild Bill Brooks. This wild name-calling is to indicate that the PV source circuits can have currents and voltages, determined by the climate. We can have voltages and currents that are greater than PV module STC (Standard Test Conditions), so for instance, we can have irradiance over 1000W per square meter, which means that the current we calculate in 690.8(A) is based on short circuit current multiplied by a correction factor of 125%. Additionally, the voltage for wild PV is calculated in 690.7 and is greater in colder places. Once electronics get involved and have the ability to limit the currents and voltages, we no longer have wild PV and the extra correction factors associated with it, and then we can potentially carry more power on a conductor, among other things.

One more thing we see in the 2023 NEC Informational Note Figure 690.1 when we compare it to the 2020 NEC, is that previously we had dc-to-dc converter source circuits and dc-to-dc converter output circuits, where the source circuits went into the dc combiner and the output circuits came out of the dc combiner after the parallel connections. Using a dc-to-dc converter dc combiner is typical of larger commercial products. Now, as of the 2023 NEC, we call what was formerly a dc-to-dc converter source circuit and a dc-to-dc converter output circuit a dc-to-dc converter circuit. We have essentially got rid of calling things output circuits inside the PV system in Article 690 of the 2023 NEC. Everything is now a source. However, on the output of a PV system, it is possible to have a PV system output circuit. This is for consistency with other source output circuits such as wind system output circuit or fuel cell system output circuit.

To sum it up all dc PV circuits, can be called PV source circuits and when a PV circuit is made up of modules in series, it can also be called a PV string circuit. All dc-to-dc converter circuits are now called dc-to-dc converter circuits.

Central Inverters, String Inverters, and Microinverters

These terms are not defined in the NEC but are commonly used in the industry.

A microinverter is usually an inverter connected directly to one or more PV modules, which is placed under the PV modules. The Inflation Reduction Act (IRA) has provisions for manufacturing incentives where a microinverter is defined as a single PV module per inverter. One rather popular brand of "microinverters" has 4 PV modules per inverter. There may be a renaissance of multiple PV modules in series per microinverter or dc-to-dc converter with the UL 3741 PV Hazard Control System (PVHCS) listing for rapid shutdown (see 690.12).

A string inverter is the term used for an inverter that has **PV string circuits** that go straight to the inverter, so no dc combiners are used. There are some huge gigawatt-scale PV projects that use string inverters.

Central inverters are large inverters where you have dc combiners in the field, combining **PV string circuits**, and then out of the combiners come **PV source circuits** (formerly known as PV output circuits), which go to the inverter.

In the previous versions of this book and the NEC there were more images of PV systems and components, which were moved to article 705, which is covered in Chapter 9. You may recall the images with ac-coupled and dc-coupled systems.

690.4 General Requirements

Outline of 690.4 General Requirements:

690.4 General Requirements
- 690.4(A) Photovoltaic Systems
- 690.4(B) Equipment
- 690.4(C) Qualified Personnel
- 690.4(D) Multiple PV Systems
- 690.4(E) Locations Not Permitted
- 690.4(F) Electronic Power Converters in Not Readily Accessible Locations

One of the more difficult things for someone learning to use the NEC is to remember to know where to look for something. This book is going to do its best to outline, organize, point to, and discuss topics, so that the reader will be more familiar with and have a better idea of where to find what they are looking for.

Section 690.4 General Requirements, which is in Part I General of Article 690, is not very memorable, and it is going to stump a few people who are looking for this information, so let us state the obvious and dive into these General Requirements. If you can make something catch your attention for whatever reason, it will help you remember it.

690.4(A) Photovoltaic Systems

In plain English: PV systems can supply a building at the same time as other sources of power.

Discussion: If you live in coal country and the AHJ refuses to let you put those fresh sunbeam electrons into the grid, you can support your argument here. North Dakota is the worst state for solar by the way. Ranked 52 out of 50 states.

690.4(B) Equipment

Equipment that needs to be listed (or field labeled) for PV applications according to 690.4(B):

Electronic Power Converters (Includes inverters and dc-to-dc converters)
Inverters (UL 1741) (UL 1741SA in California)
Motor Generators (Dc motors driving a rotating generator.)
PV Modules (UL 1703 old or UL 61730 newer)
PV Panels (Products have been built that panelize modules and have been listed when shipped that way. Solyndra solar panels were a group of tubes, each being a module, which were listed and shipped with multiple tubes on a rack creating a solar panel and listed to UL 1703.)
Ac Modules (UL 1703 or UL 61730 and UL 1741 tested as a unit.)
Dc Combiners (UL 1741)
Dc-to-Dc Converters (UL 1741)
Charge Controllers (UL 1741)
PV Rapid Shutdown Equipment (PVRSE) (UL 1741)
PV Hazard Control Equipment or System (PVHCE or PVHCS) (UL 3741)
Dc Circuit Controllers (UL 1741)

> Discussion of listed and field labeled equipment: Listed products are found on a list of certified products that various certification labs develop. Field labeled products may not be on one of these lists but get evaluated by a certification lab which puts a label on the product after it has met whatever test requirement was requested to be tested. For example, say that your 5MW inverter was not listed to UL 1741 and was made in Europe. A testing lab can come out and test and field label the inverter to certify that it is compliant with the project specifications. This can be expensive and most often avoided by using equipment listed to UL 1741.

690.4(C) Qualified Personnel

What it means: Installation of equipment and wiring should be done by "qualified personnel."

There is an **informational note** that tells us that we can look to Article 100 Definitions to see the definition of **qualified person**. Article 100 **Qualified Person Definition**: One who has skills and knowledge related to the construction and operation of the electrical equipment and installations and has received safety training to recognize and avoid the hazards involved.

Discussion: Some would say that a qualified solar installer is NABCEP PV Installation Professional (PVIP) Certified. Others would say only an electrician should install solar and yet others say only a roofer should put a hole in a roof, or only a pile driver should put a pile in the ground. There is also a UL Solar Certification. Where are all the UL Certified Installers? (There is one of the few writing this book.) Perhaps reading this book and taking our classes makes you qualified or is at least part of the process. You are more qualified than you were two pages ago. Perhaps this "qualified" definition is purposely vague, like many things in the NEC, so that different AHJs can have different interpretations to fit their environmental and political climates.

An Informational Note on Informational Notes

Informational notes in the NEC are good ideas, but not requirements. Just like a yellow speed sign tells you it is a good idea to slow down for a corner, an informational note gives you good advice. Informational notes used to be called fine print notes and abbreviated FPN. If you read the NEC Handbook, rather than the Codebook, there is more commentary in the Handbook, which is not part of the Code, but tries to help. Sometimes the commentary disagrees with the Code, which is ironic.

690.4(D) Multiple PV Systems

What it means: multiple PV systems are allowed on a single building.

If multiple PV systems on a building are located away from each other, then there must be a directory at each PV system disconnecting means showing where the other disconnecting means are located in accordance with **705.10 Directory**.

Discussion: We do not want firefighters thinking they turned off all the PV on the building when they hit one of the disconnects on the building, not knowing that there are other disconnects that will turn off other PV systems at different locations on the building. Tricking firefighters or utility workers is not cool nor is it allowed.

Disconnecting Means Means …

As you would like to think, a PV system disconnecting means is an off switch for a PV system. A disconnecting means is what separates a PV system from the rest of the electrical system. A PV system disconnecting means for an interactive (grid-tied) inverter would be on the ac side of the inverter, separating the PV system from what is not the PV system.

We generally have one PV system disconnecting means and several equipment disconnects for a PV system. Some rare and complicated dc PV systems could have more than one disconnect to form the PV system disconnecting means, but these disconnects have to be grouped for each system.

Study PV System and Source Disconnects in Figures 9.1 and 9.2 in the Article 705 images in Chapter 9 on pages 190 and 191 of this book. These images were formerly in Article 690 in earlier versions of the NEC.

690.4(E) Locations Not Permitted

PV equipment and disconnecting means are **not allowed in bathrooms**, just in case you had your heart set on mounting one next to your toilet—sorry, not allowed. Put that TP roll somewhere else than the handle of the dc disconnect.

Think of "wet feet" and getting shocked.

690.4(F) Electronic Power Converters Mounted in Not Readily Accessible Locations

Electronic power converters (inverters, dc-to-dc converters, etc.) can be on roofs or other exterior areas that are not readily accessible. This clearly tells us among other things, that there is no problem putting module level power electronics (MLPE) underneath PV modules.

690.4(G) PV Equipment Floating on Bodies of Water (New and as exciting as a boat trip!)

Here is the exact and easily understandable wording:

> PV equipment floating on or attached to structures floating on bodies of water shall be identified as being suitable for the purpose and shall utilize wiring methods that allow for any expected movement of the equipment.
>
> Informational Note:
> PV equipment in these installations is often subject to increased levels of humidity, corrosion, and mechanical and structural

stresses. Expected movement of floating PV arrays is often included in the structural design.

Discussion: There was an effort to create a required UL listing for floating equipment, but no standard has been written yet, so it is not in the NEC. Identified means that the equipment can be used in wet and/or salty conditions that may have some wave action. Water and electricity together improperly can be dangerous. The specific requirements here are open to interpretation and probably an engineer familiar with marinas would be helpful.

We like to call floating photovoltaics phloat-o-voltaics—a term invented 15 years ago by a company named SPG Solar!

690.6 Alternating-Current (Ac) Modules

Outline of 690.6 Alternating-Current (Ac) Modules:
Alternating-Current (Ac) Modules
- A PV Source Circuits
- B Output Circuit

Discussion: 690.6 is stating the obvious.

690.6(A) PV Source Circuits

What it means: ac modules are tested and listed as a unit, so we do not need to consider any dc circuits, such as PV source circuits.

It is interesting to note that, with a microinverter, we consider the dc conductors between the module and the inverter a PV source circuit, but not with an ac module.

690.6(B) Output Circuit

It says: The output of an ac module is considered an inverter output circuit.

Discussion: This is obvious but needs to be explained in case an AHJ gives you a problem. A question PV professionals often have is: "What is the difference between an ac module and a microinverter bolted to a PV module?" The answer is that if the PV module was listed to UL 1703 or UL 61730 while the inverter was bolted to it, and if the inverter was tested and listed to UL

1741 while bolted to the PV module, then it is an ac module and we do not consider the dc part of the product when installing this module.

AC PV Module UL Testing

If the module and microinverter were not listed together, then we are responsible for applying the NEC to the dc circuit, going from the module to the inverter. It is also interesting to note that the word "microinverter" does not appear in the NEC. The NEC looks at a microinverter as nothing more than a small (micro) inverter.

Certification Listings for PV Modules

The listing that we have been used to for years for PV modules was UL 1703, however in 2020 there was a new listing, which is 61730. The main reason for this change, is so that one single test can be used globally. Before, if we wanted to use the same PV module in both Europe and the US, it would have to go through at least two sets of tests: the CE test for Europe, and the UL test for the US. If you look closely at the number, you can see that if you put a 6 before 1703 and then switch 0 and 3, you get 61730. The 6 in front of the listing is for IEC (International Electrotechnical Commission) and 61703 was taken, hence we have 61730.

If a PV module was first designed before December 2020, even if it is still being made, then you can still use the UL 1703 listing. If it was a new product after December 2020, then it should have the UL 61730 listing.

Requiring fewer mostly redundant tests for a global PV module helps bring down the price of PV.

End of 690 Part I General

2 Article 690 Photovoltaic Systems Part II Circuit Requirements

Part II Circuit Requirements:

690.7 Maximum Voltage
690.8 Circuit Sizing and Current
690.9 Overcurrent Protection [Article 240 is also Overcurrent Protection.]
690.11 Arc-Fault Circuit Protection (Direct Current)
690.12 Rapid Shutdown of PV Systems on Buildings [discussed in Chapter 3.]

690.7 Maximum Voltage

Understanding 690.7 sets true solar professionals apart from the solar un-professionals. Understanding calculations using 690.7 is also very important to NABCEP, as reflected in their exams; however, the Module Level Power Electronics (MLPE) are reducing the need for string sizers, especially for those designing PV systems on buildings.

Outline of 690.7

690.7 Maximum Voltage
- 690.7(1) PV dc circuits not to exceed 1000V on buildings
- 690.7(2) PV dc circuits not to exceed 600V on 1- and 2-family dwellings
- 690.7(3) PV circuits exceeding 1000V on buildings shall comply with 690.31(G)

DOI: 10.4324/9781003189862-3

- 690.7(A) Photovoltaic Source Circuits
  - 690.7(A)(1) Calculations with low temperature and Voc coefficients
  - 690.7(A)(2) Using Table 690.7(A) Voc Correction Factors for Silicon Crystal Modules
  - 690.7(A)(3) PV Systems Over 100kWac Under Professional Engineer (PE) Supervision
- 690.7(B) Dc-to-Dc Converter Circuits
  - 690.7(B)(1) Single Dc-to-Dc Converter
  - 690.7(B)(2) Two or More Series Connected Dc-to-Dc Converters
- 690.7(C) Bipolar PV Source Circuits
- 690.7(D) Marking DC PV Circuits

Electricians are accustomed to having the grid or a factory-set device provide a reference voltage for calculations. With PV, calculations have a lot more variables.

690.7 Maximum Voltage

There are different ways of defining and calculating voltage, which we will describe below. The maximum voltage shall be the highest voltage between any two conductors of a circuit, or any conductor and ground.

690.7(1) PV dc circuits not to exceed 1000V on buildings

Discussion: PV dc circuits on buildings other than those described below in 690.7(2) and 690.7(3) cannot exceed 1000V.

690.7(2) PV dc circuits not to exceed 600V on 1- and 2-family dwellings

Discussion: PV dc circuits on 1- and 2-family dwellings cannot be over 600V.

690.7(3) PV circuits exceeding 1000V on buildings shall comply with 690.31(G) (page 122)

Analysis: This is new in the 2023 NEC. 690.31(G), which we will cover in detail later, allows limited dc PV circuits to run along the sides of buildings, so now you can have an inverter between 1000V and 1500V on the ground wired to an inverter that is mounted on the side of a building! This is common sense going into the NEC, even when it comes to rapid shutdown. Stay tuned.

690.7(A) Photovoltaic Source Circuits

PV source circuits (which include PV string circuits) get their voltage directly from series connected solar cells. The NEC will consider two factors that increase PV source circuit voltage. First, putting modules in **series increases the voltage**. Secondly, **cold temperature increases the voltage.**

690.7(A) Informational Note

An **informational note is a good idea**, not a requirement. The NEC tells us that a good place to find cold temperature data that we can use in determining voltage for locations in the United States is the ASHRAE Handbook. A very convenient place to find this data is at the website for the Expedited Permit Process: www.solarabcs.org/permitting.

What is Cold?

The Solar America Board for Codes and Standards website for the Expedited Permit Process is a document that was put together by Bill Brooks under contract of the United States Department of Energy (USDOE). On the left side of the www.solarabcs.org webpage, click on Expedited Permit Process and then click on "map of solar reference points" to find the low temperature data to use for calculating voltage. This webpage also has high temperature data that can be used for wire sizing, which we will cover later in this book.

The Expedited Permit Process is a template, which includes fill-in forms that can be used to put together a permit package. Regardless of whether you use the templates, there is a lot of good information to study by downloading the 82-page Expedited Permit Process "full report." Anyone in the solar industry will benefit from becoming familiar with this report. It also helps when studying for the NABCEP PV Installation Professional exam. There is another new version of this permitting process funded by the USDOE, called the Simplified Permit Process, that Bill put together at www.solsmart.org/permitting.

The temperature that we use is called the ASHRAE extreme annual mean minimum design dry bulb temperature. This means that half of the years on record, the temperature will get colder. Dry bulb gives us the ambient temperature, whereas a wet bulb will be cooler because of evaporation, except in 100% humidity. Remember that this is not the record low temperature, which was commonly used in the past. There are a few reasons why we do not need to worry about this record low, including the following.

(1) If the sun is shining on the PV, then the PV will be warmer than ambient.
(2) The irradiance on a cold PV module is usually much lower than 1000 W/m^2.
(3) PV module degradation will lower the voltage.
(4) If the inverter is on, then it will be operating at Vmp, which is significantly lower than Voc.

The 2023 NEC, like the 2020 NEC, gives us three ways to determine voltage and we can make a choice of which method we will use. These methods will result in different values of voltage, depending on the method we chose to use.

The **three methods** for determining PV source circuit **(string) voltage** are:

690.7(A)(1) Calculations
690.7(A)(2) Table 690.7(a)
690.7(A)(3) Engineering Supervision

690.7 (A)(1) Voltage Temperature Calculation Method

The 690.7(A)(1) method is the most common method used by solar professionals for determining PV source circuit (string) voltage. This method is also required for anyone taking any NABCEP PV certification exam.

In order to calculate the module maximum voltage, you will need three things:

(1) Low temperature (ASHRAE value)
(2) Temperature coefficient of Voc for the PV module
(3) Rated PV Module Voc (open-circuit voltage at STC)

Module **Voc** and **temperature coefficient of Voc** is most commonly found on the PV module manufacturer's datasheet. Low temperature data is most easily found at www.solarabcs.org/permitting.

Let us run through a **PV source circuit maximum voltage calculation** using a simple example with round numbers.

Example:

(1) Cold Temp = −5°C
(2) Temp. Coef Voc = −0.3%/°C
(3) Voc = 40V
(4) Number of modules in series = 10

Calculation:

(1) Determine **delta T** (difference in temperature) from Standard Testing Conditions (STC).
   a. All PV modules are tested at STC = 25°C
   b. The difference between −5°C and 25°C is **30°C or −30°C**.
(2) Multiply delta T by Temp. Coef. Voc.
   a. 30°C × 0.3%/°C = 9% increase in voltage
   b. Another easier method converts percent to decimal first.
      1. 30°C × 0.003 = 0.09
(3) Add 1 to figure above to get 109% increase in voltage
   a. 0.09 + 1 = 1.09

This figure is a temperature correction factor.

(4) Multiply the temperature correction factor by Voc at STC to get cold temperature Voc.
   a. 1.09 × 40V = **43.6V = maximum voltage** for one module
(5) 10 in series × 43.6V = **436V maximum voltage for the PV source circuit** (string)

When practiced, the method above can be done in 10 seconds by fast calculator users. If you practice this 10 times fast, you will be an expert. This method can be done easily with a calculator and without writing anything down. On the calculator keypad, press:

25 + 5 = 30 (if the 5°C were above 0 then subtract 5 from 25 to get 20)
30 × .003 =.09
.09 + 1 = 1.09
1.09 × 40 = 43.6V = maximum voltage for one module

Often, we do **string sizing** with this number, which means we divide it into the inverter maximum input voltage and then round down to get the **maximum number of modules in series** without going over voltage.

Example using 43.6V maximum voltage and 450V inverter:

450V/43.6V = 10.3

10 modules is the maximum number in series without going over voltage (always round down here). In this example, if we have 10 in series, then the maximum system voltage is:

10 in series × 43.6V = **436V = maximum system voltage**

It is very common for solar un-professionals to incorrectly write that the maximum system voltage is 450V on the label in this example, which is incorrect. 436V is correct.

Once practiced, you should be able to do this calculation without paper using a calculator in less than a minute. The world record is 14.1 seconds (3 seconds faster than the previous edition of this book and 14 seconds slower than the Rain Man method).

You do not use this method for dc-to-dc converters (optimizers), which is a common mistake. There are some module level power electronics (MLPE) which do require string sizing, so be sure to read the instructions when you are using MLPEs.

String Sizing

With string sizing calculations, and others, there are many paths to the correct answer. Try to not just memorize a formula, but to understand the steps—this way, when you make a mistake, which we all do, you will catch yourself. One trick I use with this calculation is making the delta T a positive number, just to make things easier for non-engineers. Then I just know that when it gets colder, the voltage goes up.

You can also string size for your short string, but this is an efficiency calculation and not a Code issue. The differences are that you use the Vmp, rather than the Voc, then you make the Delta T the difference between the hot solar cell temperature (about 30°C hotter than ambient hot temperature) and STC. You will take the delta T and multiply it by the temperature coefficient of power or Vmp, which is a larger number than the coefficient of Voc. The temperature coefficient of Vmp is usually not given, so the temperature coefficient of power is a good substitute. You then calculate the correction factor by subtracting rather than adding as you do for cold temperatures. For example, if your hot ambient temperature is 40C and your hot cell temperature is 70C, then your delta T is 70C – 25C = 45C. Then let's say that your temperature coefficient of Vmp or power is –0.4%/C. So, turn the percentage into a decimal by moving the decimal to the left two spaces and you get -0.004/C. Then multiply 45C × –0.004 = 0.18, which corresponds to an 18% loss. Now we are going to do subtraction for the loss of voltage, so 1 – 0.18 = 0.82 correction factor. Then we multiply the Vmp by 0.82 to get our hot temperature Vmp. We can then divide this hot temperature into our lowest voltage that we want, which can be the inverter low maximum power point tracking (MPPT) voltage. Make sure to round up this time, since we do not want to go below the operating voltage window. Usually, people want closer to the longest string, rather than the shortest string. You can also take into consideration PV module degradation here, so you do not go too low 10 years from now. It used to be that inverters did not have built-in dc-to-dc converters and had transformers, which limited the voltage window and, in some places, the longest string for the winter was sometimes too short for the hot summers.

Summary: For short string sizing, use hot temperature Vmp, which is less than Vmp, and round up to stay in the voltage window.

690.7(A)(2) Table Method for Calculating Voltage

Using Table 690.7(a) is easier than performing the **690.7(A)(1) Voltage Temperature Calculation Method**. We can consider this optional method a shortcut; however, in some cases, we will have more options for more modules in series using the 690.7(A)(1) calculation method. The **690.7(A)(2) method using Table 690.7(a)** is **more conservative** and will come up with a slightly **higher module voltage** every time.

We use **Table 690.7(a)** by cross-referencing a temperature with a **temperature correction factor**. We then multiply the temperature correction factor by the module open-circuit voltage to get the module maximum voltage.

Table 2.1 NEC Table 690.7(a) voltage correction factors for crystalline and multicrystalline silicon modules. Correction factors for ambient temperatures below 25°C (77°F). (Multiply the rated open-circuit voltage by the appropriate correction factor shown below.)

Ambient temperature (°C)	*Factor*	*Ambient temperature (°F)*
24 to 20	1.02	76 to 68
19 to 15	1.04	67 to 59
14 to 10	1.06	58 to 50
9 to 5	1.08	49 to 41
4 to 0	1.10	40 to 32
−1 to −5	1.12	31 to 23
−6 to −10	1.14	22 to 14
−11 to −15	1.16	13 to 5
−16 to −20	1.18	4 to −4
−21 to −25	1.20	−5 to −13
−26 to −30	1.21	−14 to −22
−31 to −35	1.23	−23 to −31
−36 to −40	1.25	−32 to −40

Source: Courtesy NFPA

Let us use the **690.7(A)(2) method using Table 690.7(a),** using the same numbers that we just used in the **690.7(A)(1) calculation** example; however, this time we will not use the module manufacturer's temperature coefficient for open-circuit voltage.

Example:

(1) Cold temp = −5°C
(2) Voc = 40V
(3) Number of modules in series = 10

Calculation using Table 690.7(a)

(1) Looking at Table 690.7(a) at −5°C, we can see that −5°C corresponds with a temperature correction factor of 1.12 (12% increase in voltage)
(2) Multiply 1.12 × 40V and get 44.8V
(3) 10 in series × 44.8V = 448V maximum system voltage

We can see by the results of comparing the **690.7(A)(1) calculation method** to the **690.7(A)(2) table method**, that the **690.7(a) table method** resulted in a **higher voltage** that is very close to the 450V inverter maximum voltage.

If it were 1° colder at −6°C we would still be able to have 10 in series with the **690.7(A)(1) calculation method**, but we would have gone over voltage using the **690.7(A)(2) table method** using **Table 690.7(a)**.

We can see that at −6°C in **Table 690.7(a)** we have a temperature correction factor of 1.14.

40V × 1.14 = 45.6V
45.6V × 10 in series = 456V = over voltage for our 450V inverter example

690.7(A)(3) Engineering Supervision Method for Calculating Maximum Voltage for PV Systems over 100kW Generating Capacity (Ac System Size)

Under engineering supervision, there can be alternative ways of doing things throughout the NEC. According to 690.7(A)(3), a

licensed professional electrical engineer will have to stamp the system design. A professional engineer (PE) has gone to school, worked in the field, and taken a difficult exam. The designation of PE is awarded on the state level. It is up to the AHJ to accept a stamp of a PE who is licensed in another state.

The 690.7(A)(3) Engineering Supervision Method requires that the PE uses an "industry standard method" for determining maximum voltage.

690.7(A)(3) INFORMATIONAL NOTE

There is an informational note that recommends an "**industry standard method**" for calculating maximum voltage of a PV system.

Industry Standard Method for Calculating Maximum Voltage:

Photovoltaic Array Performance Model
Sandia National Laboratory
SAND 2004-3535 www.osti.gov/servlets/purl/919131

Insiders like Bill call this the David King method, since he was an author.

To summarize the report as far as voltage goes: Taking the heating effects of irradiance into consideration, the temperature of the PV will be hotter than ambient, and we can have a lower module voltage and perhaps another two modules in series. Different solar software, such as NRELs System Advisor Model (SAM), PVsyst and Helioscope, can model using the Sandia Model.

690.7(B) Dc-to-Dc Converter Circuits

Dc-to-dc converter source and output circuits shall be calculated in accordance with 690.7(B)(1) and 690.7(B)(2).

690.7(B)(1) Single Dc-to-Dc Converter

For a single dc-to-dc converter output, the maximum voltage is the maximum rated output of the dc-to-dc converter.

Maximum voltage is what it says on the label, installation instructions, or datasheet for maximum output voltage.

Mostly we see dc-to-dc converters in series with each other in the context of the NEC, such as with SolarEdge Optimizers; however, a maximum power point tracking charge controller is essentially a dc-to-dc converter. If a charge controller can limit current coming from the battery, then it is not part of the PV system. PV systems only encompass currents coming from PV.

Dc-to-Dc Converters in Modern PV and Energy Storage Systems

As electronics mature, we are seeing dc-to-dc conversion taking place throughout the industry. These are at the inputs of our multiple MPP inverters, MPP charge controllers, and within our energy storage systems. One of the reasons that ac became our grid rather than dc, is because Tesla pioneered the technology to change ac voltage with transformers,

Figure 2.1 Tesla using improper PPE with alternating current.
Source: Courtesy Wikimedia, https://en.wikipedia.org/wiki/Nikola_Tesla#/media/File:Nikola_Tesla,_with_his_equipment_Wellcome_M0014782_-_restoration2.jpg.

which work for alternating current and not direct current. Now, with efficient modern dc conversion technology, we are seeing more applications for dc circuits. Watch out Nikola Tesla, Thomas Edison is coming for you!

690.7 B)(2) Two or More Series Connected Dc-to-Dc Converters

Maximum **voltage is determined in accordance with instructions** of the dc-to-dc converter.

If instructions are not included, sum up the voltage of the dc-to-dc converters connected in series.

Discussion: **Dc-to-dc converters can electronically limit voltage** when connected in series. There are, however, some dc-to-dc converters that, along with having the capability to convert voltages, also can bypass the converter internally and send PV voltage through the converter. Dc-to-dc converter installation instructions and help lines will be your best source of determining maximum voltage.

The most commonly used dc-to-dc converter as of the writing of this book has a maximum system voltage between 480V and 1000V, which you can find in the datasheet or installation manual.

690.7(C) Bipolar PV Source Circuits

A bipolar circuit has a positive grounded monopole circuit and a negative grounded monopole circuit. Maximum voltage is considered voltage to ground.

If there is a ground-fault or an arc-fault, the inverter is required to isolate both circuits from each other and from ground.

Discussion: From a safety point of view, the voltage of your house is limited to 120V to ground since voltage to ground relates to safety. In a similar way, a bipolar system with a maximum voltage of 1000V to ground can have a voltage of 2000V measured array to array! However, a 2000V bipolar system is allowed to use 1000V rated PV modules since the modules never see more than 1000V in this configuration. We have not seen bipolar inverters used in about a decade but expect to see them sometime between the publication of this book and the publication of our future book on the 2026 NEC. We can tell the future.

690.7(D) Marking Dc PV circuits

690 Part VI Marking is no longer in the 2023 NEC, as it was in the 2020 NEC. 690.53 DC PV Circuits (in Part VI Marking) was moved over here to 690.7(D) and there were no changes, besides the move.
Dc PV circuits require a label at one of the following locations:

(1) Dc PV system disconnecting means
(2) PV system electronic power conversion equipment
(3) Distribution equipment associated with PV system

All this sign requires now is the maximum dc voltage from 690.7.
In previous versions of the Code, we had requirements for maximum circuit current in 2017 NEC, along with operating voltage (Vmp) and current (Imp) in 2014 and earlier versions of the NEC.

There are no color, size, or reflective requirements in the NEC for the 690.53 Dc PV Circuits sign, so get creative! All it has to be is permanent and readily visible.
Summing up 2023 NEC changes in 690.7:

(1) We can have circuits over 1000V on buildings if you follow the restrictions, so you can mount your inverter for a ground mounted PV system on a building.
(2) We are getting rid of "output circuit" terminology for wild PV, dc-to-dc converters, and bipolar circuits.
(3) Dc voltage marking requirements that were in 690.53 are now in 690.7.

Most changes here and in general are organizational and/or made to clarify things. The only real 690.7 change is the part about being able to put your dc circuits over 1000V on the building, and the details and limitations for that can be seen at 690.31(G) on page 122. This does not mean that we can put a PV array over 1000V on any building, just part of the circuit. Which usually is going to an inverter that we can now hang on a wall.

Outline of 690.8 Circuit Sizing and Current

690.8(A) Calculation of Maximum Circuit Current
- 690.8(A)(1) PV System Circuits
  - 690.8(A)(1)(a) PV Source Circuit Calculation by one of the following
    - 690.8(A)(1)(a)(1) 125 Percent of Short Circuit Current
    - 690.8(A)(1)(a)(2) Systems Over 100kWac Engineering Supervision
  - 690.8(A)(1)(b) Dc-to-Dc Converter Circuit Current
  - 690.8(A)(1)(c) Inverter Output Circuit Current
- 690.8(A)(2) Circuits Connected to the Input of Electronic Power Converters

690.8(B) Conductor Ampacity
- 690.8(B)(1) Without Adjustment and Correction Factors
- 690.8(B)(2) With Adjustment and Correction Factors

690.8(C) Systems with Multiple Direct-Current Voltages

690.8(D) Multiple PV String Circuits

690.8 Circuit Sizing and Current

This is how we define current and ampacity for **wire sizing** and equipment selection. Wire sizing is not simple, so pay close attention and come back often. **690.8(A) defines what current is and 690.8(B) defines the required ampacity of the wire.**

690.8(A) and 690.8(B) Overview

Reinforce in your synapses that **690.8(A) defines currents** used for wire sizing and **690.8(B)** gives us different checks to perform to make sure that the **wire can handle the current under different conditions**, such as heat and being oversized for continuous current.

There are still **other checks** used for wire sizing regarding overcurrent protection found in Article 240 of the NEC. In **Article 240 Overcurrent Protection** we have to make sure that the **overcurrent protection** device is going to protect the wire. See wiring sizing examples in Chapter 12, the most difficult chapter in the book.

690.8(A) Calculation of Maximum Circuit Current

We will **define currents** used for wire sizing for different circuits in a PV system. PV system currents are more complicated than currents in circuits most electricians are used to dealing with. This is because we have **some circuit currents that increase with the brightness** of light and **other** circuit currents that are **limited by smart electronics**.

690.8(A)(1) PV System Circuits

690.8(A)(1) says currents are determined by 690.8(A)(1)(a) through 690.8(A)(1)(c). There was a little rearranging here in the 2023 NEC with the PV output circuit and dc-to-dc converter output circuit definitions being removed, and then combined under the headings of PV source circuits and dc-to-dc converter circuits. Currents are still the same here, it is just a different arrangement (FPN: an amp is still about 6 quintillion electrons passing by in a second).

690.8(A)(1)(a)(1) 125% OF SHORT-CIRCUIT CURRENT METHOD

This is the typical way we define maximum circuit current for "wild" PV circuits and is the way we have always done it in the past:

125% of short-circuit current
Isc × 1.25 = maximum circuit current

Discussion: This is the definition of maximum circuit current for one module, or a number of modules connected together in series. Some would argue that it is overdoing it to base your wire size on a short circuit. This is done because we are accounting for increased irradiance over STC (1000 w/m^2).

Also, it is unusual for electricians to see a short circuit that is only about 7% more than operating current, as it is with PV modules. Usually when you short something besides PV, a huge amount of current tries to rush in and then an overcurrent protection device opens the circuit.

Let's look at some of the PV module circuit currents in order of increasing current:

Imp = current at maximum power (not a code thing, but your operating current at STC)
Isc = short-circuit current
Isc × 1.25 = Maximum circuit current from applying 690.8(A)(1)(a)(1)
Isc × 1.56 = Required ampacity for continuous current from 690.8(B)(1) for wild PV

Note: Maximum circuit current (Isc × 1.25) was required on the label of dc disconnects in the 2017 NEC; however the requirement was discontinued in the 2020 NEC.

It is considered by many smart people to be overdoing it by using 125% of short-circuit current to define maximum circuit current. The reason it was not a big deal in the past is because PV was so expensive that oversizing wires was done to keep as much of our expensive energy as possible from being lost in the wires (voltage drop). Since we have entered the age of inexpensive PV energy, it makes more sense to use a smaller wire than it used to, and this leads us to a new way to define current in **690.8(A)(1)(a)(2), the Engineering Supervision Method.**

Irradiance Outside of Earth's Atmosphere

It is interesting to note that Imp × irradiance in space at Earth's orbit, which is about 1366W per square meter, is about equal to Isc × 1.25.

Imp × 1.366 = Isc × 1.25

Coincidence or conspiracy? Clearly the Martians are planning to suck away our atmosphere and want to make sure our electrical code does not need to change after our atmosphere is gone …

690.8(A)(1)(a)(2) ENGINEERING SUPERVISION METHOD FOR CALCULATING MAXIMUM CIRCUIT CURRENT FOR PV SYSTEMS OVER 100KW GENERATING CAPACITY (AC SYSTEM SIZE)

As with 690.7(A)(3) for determining voltage, 690.8(A)(1)(a)(2) allows a licensed professional electrical engineer **(PE)** to use an

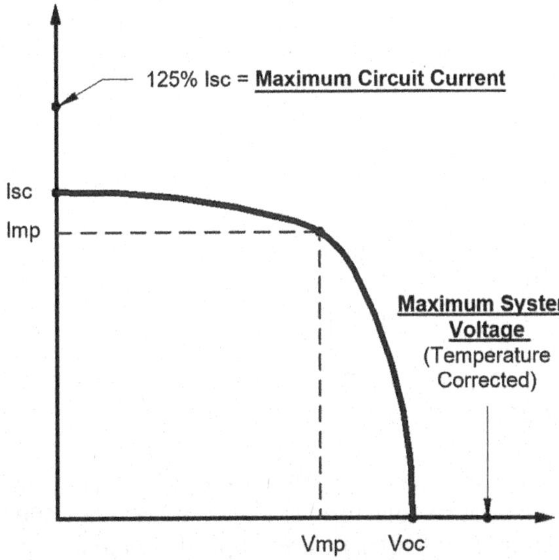

Figure 2.2 IV curve with different currents plotted showing maximum circuit current, which is used for sizing wires, above and beyond short circuit current.

Source: Courtesy Robert Price, AxisSolarDesign.com.

industry standard method. This method is based on the **highest three-hour current average** from **simulated local irradiance accounting for elevation and orientation**.

This industry standard method **must also not be less than 70% of the 125% of Isc value used in 690.8(A)(1)(a)(1)**.

Since 690.8(A)(1)(a)(1) is Isc × 1.25, then the industry standard method cannot be less than 70% of 125% of Isc, so 0.7 × 1.25 = 0.875

This means that this **industry standard method cannot be less than 87.5% of Isc**.

We could call this the **not less than 87.5% Isc method**.

This does not mean that the PE does not have to do anything besides multiply Isc × 0.875. The PE will also have to analyze the PV system, including taking irradiance, elevation, and orientation into consideration.

Higher elevation causes more current, due to less atmospheric filtering (closer to space).

THE 690.8(A)(1)(a)(2) INFORMATIONAL NOTE

The Informational Note for 690.8(A)(1)(a)(2) points the professional electrical engineer in the same direction as 690.7(A)(3) and towards the same report mentioned earlier in 690.7 for voltage engineering supervision:

Industry Standard Method for Calculating Maximum Circuit Current:

Photovoltaic Array Performance Model
Sandia National Laboratory
SAND 2004–3535 www.osti.gov/servlets/purl/919131

The 156 Factor

Current and ampacity with PV source circuits has confused many prospective solar professionals. When both **690.8(A)(1)(a)(1)** and **690.8(B)(1)** are applied simultaneously, a **resulting multiplication factor of 156%** is used. Oftentimes people try to apply this **156%** factor to circuits besides PV source circuits. This multiplication factor is **only applied** to circuits that are **directly and proportionally influenced by sunlight** and ***not*** circuits that are limited by **electronics**. We can call this "**wild PV**" when PV is not limited by electronics.

156% comes from two different 125% correction factors. The **690.8(A)(1)(a)(1)** correction factor is for **natural irradiance beyond the standard testing conditions of 1000W per square meter**, which is how all PV modules are tested and rated. The other 125% in **690.8(B)(1)** is the **required ampacity for continuous current** that we have for all our solar circuits. **Continuous current** as defined in the NEC is a current that can last over three hours. PV can last all day—particularly in tracking systems! Required ampacity for continuous current is not a current, it is more like a safety factor, and a way to get you to buy more copper or aluminum.

As a side note, did you know that many silicon solar cells are 156mm × 156mm? Coincidence or photovoltaic numerologist conspiracy? See page 156 and read it backwards (kidding).

This 156% of Isc factor was a 690.8(A) informational note that was removed in the 2020 NEC because everyone read the first edition of this book and no longer needed the information.

690.8(A)(1)(b) DC-TO-DC CONVERTER CIRCUIT CURRENT

The maximum current shall be the sum of the parallel connected continuous current output rating.

Discussion: Dc-to-dc converters can electronically control current and voltage. In the instructions of the dc-to-dc converter there should be details on what the maximum current of the circuit can be. One of the benefits of the dc-to-dc converter is its smart ability to limit current, so that smaller wires can be used. For the most commonly used dc-to-dc converter (initials SE) at the time of this writing, the current of a dc-to-dc converter source circuit is 15A for some products. Converter source circuit currents are rising with 18A and even higher currents becoming common. It is also humorous that they tell us to use a wire size that is at least 11 AWG. Has anyone ever seen an 11 AWG wire? At least we don't need to have a greater-than or equal-to sign there.

Combiners Combining Combiners

In many utility scale systems, combined PV circuits will be combined again in other dc combiners, which are sometimes called "**recombiners.**" The IEC (International Electrotechnical Commission) calls what the NEC calls a "**dc combiner**" an "array junction box" if there is a single level of combining and if there are two levels of combining, then what we call the "**dc combiner**" in the NEC is called the "**sub-array junction box**" and the next level of combining is called the "**array junction box.**"

The NEC term for a combiner box is **dc combiner**.

690.8(A)(1)(c) INVERTER OUTPUT CIRCUIT CURRENT

The maximum current shall be the inverter **continuous** current output **rating**.

Discussion: The current is often marked on the inverter. If current is not marked on the inverter, you can **calculate current by dividing power by operating voltage**. It is interesting to note that dividing power by voltage does not always get the exact same value for current as the inverter datasheet. This can be due to changes in power factor (current and voltage being out of phase with each other and causing a higher current than power divided by voltage would indicate). This is one reason we see equipment such as transformers rated in apparent power, such as kVA, rather than real power, such as kW.

690.8(A)(2) Circuits Connected to the Input of Electronic Power Converters

Protecting circuits the old fashioned way (was new in 2020 NEC).

NEC wording: Where a circuit is protected with an overcurrent device not exceeding the conductor ampacity, the maximum current shall be permitted to be the rated input current of the electronic power converter input to which it is connected.

What this means: We have always based the conductor size (current) of the PV dc circuits based on the rating of the currents coming from the direction of the PV (think Isc × 1.25). Now we are allowed another way to size these circuits based on the maximum current that the ***electronic power converter*** can accept. An electronic power converter is usually the input of an interactive inverter.

This is very much like sizing the input of a battery inverter. We do not size a battery inverter input circuit based on the short circuit rating of the battery, which could be many thousands of amps. We will put an overcurrent device between the battery and the battery inverter and then size the wire based on the maximum current that the battery inverter will accept. This is also how many circuits are sized, such as loads on the grid. We do not size the loads based on the short circuit rating of the grid, we size load circuits based on how much current the load can take and then put an overcurrent device on the circuit, to protect from overcurrents and short circuits. As PV goes down in price, we are more likely to oversize the PV array and invoke 690.8(A)(2).

Summary: We can also size electronic power converter input circuits based on the maximum input current if the circuit is protected by an OCPD (overcurrent protection device).

690.8(B) Conductor Ampacity (Ability to Carry Current)

Conductor Ampacity Code References (Abbreviated and Interpreted)

690.8(A) PV Circuit Current Definitions

690.8(A)(1)(a)(1) PV Source Circuit Current Definition = Usually Isc × 1.25

690.8(B)(1) Before Application of Adjustment and Correction Factors

690.8(B)(2) After the Application of Adjustment Factors

Table 310.15(B)(1)(1) Ambient Temp. Correction Factors Based on 30°C

Formerly Table 310.15(B)(1) in 2020 NEC and Table 310.15(B)(2)(a) in 2017 NEC

Table 310.15(C)(1) Adjustment Factors for More than Three Current-Carrying Conductors

Formerly Table 310.15(B)(3)(a) in 2017 NEC

Table 310.16 Ampacities of Insulated Conductors *not* in Free Air

Formerly Table 310.15(B)(16) in 2017 NEC

Table 310.17 Ampacities of Insulated Conductors in Free Air

Formerly Table 310.15(B)(17) in 2017 NEC

Note: See Chapter 12 "PV Wire Sizing Examples" for example calculations. These calculations are nothing new to the NEC and seen throughout the different articles, including Article 706 Energy Storage Systems, which we cover in Chapter 10.

Attention! The following is an important part of 690.8(B) that needs to be understood and is commonly misunderstood:

Word-for-word NEC:

690.8(B) Conductor Ampacity: ***Circuit conductors shall have an ampacity not less than the larger of 690.8(B)(1) or (B)(2)***

The text in bold above needs to be properly understood before going further. We will do a check for 690.8(B)(1) and we will also do a check for 690.8(B)(2), and then we will choose the larger wire of the two checks. The wire will always be able to carry as much or more current than the device it is connected to, to be on the safe side. There will also be other checks. (See **Chapter 12 Wire Sizing** for examples).

Ampacity = Current-carrying ability

690.8(B)(1) Without Adjustment and Correction Factors

Here we account for **required ampacity for continuous current** by **multiplying the currents we defined in 690.8(A) by 1.25.**

Or said more simply:

690.8(A) defined current × 1.25 = 690.8(B)(1) required ampacity.

Some people call this 690.8(B)(1) value "current." **It is not really current**; it is a required ampacity that is more than the actual current in order to be extra safe due to current lasting three hours or more (continuous current).

690.8(B)(1) = Required Ampacity for Continuous Current

About Continuous Currents

PV system currents are continuous. Throughout the Code, we size wires much like we do in Article 690, and since our earth spins slowly (it takes a whole day!), the sunlight can last three hours or more, making PV currents subject to the 1.25 correction factor for continuous currents. This is where the 690.8(B)(1) correction factor comes from.

Terminal announcement regarding 690.8(B)(1) value!

Terminal = what the end (terminal) of a wire is connected to. Example: screw terminal. Why do we take terminal temperature limits into consideration?

If a wire is connected to a terminal, there can be some resistance at the connection and the terminal can heat up. The wire can act as a heat sink absorbing heat from the hot terminal. Additionally, if the wire heats up and is connected to the terminal, the terminal will also become hot.

If you open up your NEC to Table 310.16 or 310.17, you will see values for ampacity for a particular wire that change as you go across the table, depending on the **temperature rating of the insulation** of the conductor. We see 60°C ampacity on the left next to 75°C ampacity and then 90°C ampacity. A conductor that has **an insulation rating that can be hotter will be able to carry more current.**

Terminals have temperature limits, and **terminals that are 75°C rated and used with 90°C rated wires are most common in the solar industry**. In general, when doing the **690.8(B)(1)** check, we **use the 75°C column** in the **310.16 and 17** tables in this case.

If we were using 90°C rated terminals with 75°C rated wire, we would use the 75°C column, since it is less than 90°C. It is so uncommon to purposely use 90°C terminals with 75°C wire that the only time it is done is when someone is making a mistake. **90°C terminals are rare on both ends of a circuit**. Both ends of the circuit must have 90°C rated equipment to use the 90°C ampacity at the terminals. This is still done sometimes. *90°C* terminals usually require compression lugs, rather than screw terminals and compression lugs can require a hydraulic crimper.

On another note, the NEC tells us that if the terminal temperatures are not indicated, we should assume 60°C terminals if the terminals are rated for 100A or less. This situation of assuming 60°C terminals is probably only seen on an exam. The vast majority of PV equipment has 75°C or 90°C terminals.

This terminal temperature logic is only used for the 690.8(B)(1) check and not for the 690.8(B)(2) check. Repeat for emphasis: **do not use terminal temperatures or the 125% correction when doing the 690.8(B)(2) check, only for 690.8(B)(1).**

Terminal Temperature NEC Reference

Article 110 Requirements for Electrical Installations:

110.14 Electrical Connections
110.14(C) Temperature Limitations

110.14(C) reads: "The temperature rating associated with the ampacity of a conductor shall be selected and coordinated so as **not to exceed the lowest temperature rating of any connected termination**, conductor, or device. Conductors with **temperature ratings higher than specified for terminations** shall be permitted to be used for **ampacity adjustment**, correction, or both."

Essentially, what this is saying is that the terminal is rated for a certain temperature (75°C for instance). The conductor can reach 75°C, but is not permitted to get hotter than 75°C. When we apply conditions of use [adjustment and correction factors from 690.8(B)(2)], we are doing calculations that approximate how much current it takes to get the copper to 75°C. We don't do anything to the terminal. It's just reflecting the conductor temperature.

If we analyze the last sentence in 110.14(C)—

> Conductors with **temperature ratings higher than specified for terminations** shall be permitted to be used for **ampacity adjustment**, correction, or both.

—we see that we do not need to apply the terminal temperature rating when we are **applying adjustment factors**, which we will read about next.

690.8(B)(2) With Adjustment and Correction Factors

The conductor ampacity should be able to handle the currents as defined in 690.8(A)(1) after the application of correction and **adjustment factors.**

Here are the correction and adjustment factors in the 2023 NEC:

(1) Table 310.15**(B)(1)(1)** Ambient Temp. **Correction Factors** Based on 30°C
(2) Table 310.15**(C)(1) Adjustment Factors** for More than Three Current-Carrying Conductors

Additional adjustment in 2014 NEC, but not in later NECs:

(1) Table 310.15(B)(3)(c) Ambient Temperature Adjustment for Raceways or Cables Exposed to Sunlight on or Above Rooftops **no longer exists.**
   - In 2023 NEC according to 310.15(B)(2), if Raceway is ¾ inch or less above the roof, we still add 33°C to ambient temperature. We advise never putting your wiring so close to the roof. This was reduced from ⅞ inch in the 2020 NEC. Besides being hotter closer to the roof, being that close does not allow space for debris to go under the conduit.

These 690.8(B)(2) **adjustment factors** are also commonly called "**conditions of use**" since the adjustments have to do with wires put in areas where there will be more heat as with **Table 310.15(B)(1)(1) Ambient Temp. Correction Factors Based on 30°C** or where the wires will have less of an ability to dissipate heat as with **Table 310.15(C)(1) Adjustment Factors for More than Three Current-Carrying Conductors.**

Remember that when we apply these 690.8(B)(2) correction and adjustment factors, we ***do not* apply the criteria in 690.8(B)(1) such as the 125% continuous current calculation or the terminal temperature limits.**

TABLE 310.15(B)(1)(1) AMBIENT TEMP. CORRECTION FACTORS BASED ON 30°C

This table is used to correct (some say derate) the ampacity of conductors in Tables 310.16 and 310.17 for temperatures that are different from 30°C. Tables 310.16 and 310.17 are based on temperatures of 30°C, which makes the tables all an electrician has to use if working inside of a building that is not a sauna. When we are working outside of a building in the sun, temperatures can get hotter than 30°C and we can compensate for this by multiplying the conductor ampacity by the derating factor in Table 310.15(B)(1)(1).

What is Hot?

A good place to find high ambient temperatures to use for wire sizing is the www.solarabcs.org/permitting Expedited Permit Process map of solar reference points where we also found the cold temperatures that we used for calculating voltage in 690.7. It is recommended to use the temperature value given as the ASHRAE 2% high temperature for wire sizing (no need to use the higher 0.4% high temperature). This is not a record high temperature, but a temperature agreed on by many industry experts and copper barons (the Copper Development Association).

310.15(C)(1) ADJUSTMENT FACTORS FOR MORE THAN THREE CURRENT-CARRYING CONDUCTORS

If there are more than three current-carrying wires together in a raceway or cable, then we use the adjustment factors in Table 310.15(C)(1). The reason we do this adjustment is that the extra conductors in a tight space will generate more heat that has to be dissipated.

We **do *not* count a neutral** that is only **carrying unbalanced currents** from other conductors in the same circuit as a current-carrying conductor. We also **do *not* count equipment grounding** conductors since they do not carry current.

TABLE 310.15(B)(3)(C) AMBIENT TEMPERATURE ADJUSTMENT FOR RACEWAYS OR CABLES EXPOSED TO SUNLIGHT ON OR ABOVE ROOFTOPS (LAST SEEN IN 2014 NEC)

In the 2014, 2011, and 2008 NEC according to Table 310.15(B)(3)(c), we had to add to the ambient temperature a temperature adder if there were conductors in a raceway exposed to sunlight over a rooftop. The reasoning behind this temperature adder is that a conduit in sunlight can act like a solar thermal heater for wires. Apparently, due to global cooling, it was not a serious enough consideration to leave the table in the 2017, 2020, and 2023 NECs.

The 2023 NEC still tells us to add 33°C to raceways or cables that are less than 3/4 inch above the roof in sunlight. We recommend installing all conductors more than 1 inch above a roof for other reasons besides heat. **Wiring methods close to the roof encourage debris build-up**, which can cause many other problems, such as **roof rot**.

To sum up these **application of adjustment factors**, we take the 690.8(A)(1) **or** 690.8(A)(2) defined currents and apply the adjustment factors to determine if the conductor is able to carry the current. Oftentimes with rooftop PV it is this adjustment factor method that is the weak link and determines the wire size. Recall that the brand new **690.8(A)(2) currents are defined by the rated input current of electronic power converters.**

We will cover these tables in more detail in our wire-sizing example chapter at the end of the book in Chapter 12 on page 287.

690.8(C) Systems with Multiple Direct-Current Voltages

If a PV power source has **multiple output circuits** with **multiple output voltages** and employs a common return conductor, the ampacity of the **common return conductor** shall not be less than the sum of the ampere ratings of the overcurrent devices of the individual output circuits.

If we use a single common return wire for three circuits, the return wire must be capable of handling the current of all three circuits. We used to see this done with 48-volt PV systems connected to batteries.

690.8(C) is rarely if ever used.

690.8(D) Multiple PV String Circuits

Word-for-word NEC:

> Where an **overcurrent device** is used to protect more than one set of **parallel-connected PV string circuits**, the **ampacity of each conductor protected by the device** shall **not be less than** the **sum** of the following:

690.8(D)(1) The rating of the overcurrent protection device.

690.8(D)(2) The sum of the maximum currents as calculated in 690.8(A)(1)(a) for other parallel-connected PV string circuits protected by the overcurrent device.

What we are talking about here is common with low current thin film PV modules, where you can have multiple strings of modules in parallel protected by a single overcurrent protection device, such as an inline fuse inside of a wiring harness.

So, the ampacity of each conductor, including those going to each PV module, shall not be less than: (the rating of the OCPD) + (the rating of other parallel connected strings).

Say for instance you have five different 1A strings (1A after the 690.8(A)(1)(a)(1) 125% calculation) paralleled together before being protected by a single 10A fuse. If there were a short circuit on one of the conductors on the side of the fuse with the 5 strings, then there are two different sources of current. One source is from the other 4 strings and the other source is backfeeding from the rest of the array through the 10A fuse. This gives us a possibility of 14A that the conductor would have to be able to handle, so we would need a conductor that was sized to carry at least 14A. Additionally, the PV module maximum series fuse rating should be at least 14A.

The PV module parallel-connected circuit shown in Figure 2.3 below is most often used with low current thin film PV modules in utility scale projects using central inverters.

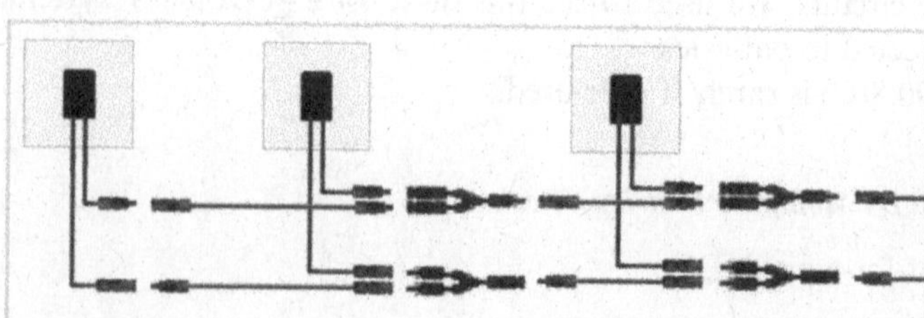

Figure 2.3 Module interconnect for multiple parallel-connected module circuits.
Source: Courtesy Shoals.

Discussion: Some PV systems have modules with low current and higher voltage. These modules are sometimes connected in parallel to a single overcurrent protection device.

If this 690.8(D) discussion was too much for you, join a large club and turn the page. This is only relevant to a small fraction of a percentage of people designing PV systems using First Solar modules. For most of us, just remember that usually greater than two in parallel requires fusing.

690.9 Overcurrent Protection

690.9(A) Circuits and Equipment
- 690.9(A)(1) Circuits Where Overcurrent Protection Not Required
  - 690.9(A)(1)(1) Conductors Have Sufficient Ampacity
  - 690.9(A)(1)(2) Currents Do Not Exceed Maximum Overcurrent Protective Device Specified
- 690.9(A)(2) Circuits Where Overcurrent Protection is Required on One End
- 690.9(A)(3) Other Circuits
  - 690.9(A)(3)(1) Conductors Less Than 10 ft Outside Protected on One End
  - 690.9(A)(3)(2) Conductors Less Than 10 ft Inside Protected on One End
  - 690.9(A)(3)(3) Conductors Protected on Both Ends
  - 690.9(A)(3)(4) Conductors Outside Protected at One End and a. Through e.
    - 690.9(A)(3)(4)(a) Installed in Metal, Underground or in Pad
    - 690.9(A)(3)(4)(b) Terminate on Single Breaker or Set of Fuses
    - 690.9(A)(3)(4)(c) OCPD Within or Within 10 ft of Disconnect
    - 690.9(A)(3)(4)(d) Installed Outside or Within 10 ft of Entrance Inside

690.9(B) Device Ratings
- 690.9(B)(1) Not less than 125% of Maximum Currents
- 690.9(B)(2) 100% of its Rating

690.9(C) PV System Dc Circuits

690.9(D) Transformers

690.9(D) Exception: Permitted Without OCPD on Inverter Side

Article 690.9 *PV Overcurrent Protection* follows the line with **Article 240 Overcurrent Protection**, but with special provisions for PV that are different from most electricity, such as solar cells that produce current based on the brightness of light, the current-limited aspects of PV, and current that can be flowing in different directions.

690.9(A) Circuits and Equipment

"PV system dc circuit and ac inverter output conductors and equipment shall be protected against overcurrent."

Discussion regarding above sentence: Notice how the first sentence in 690.9(A) does not say conductors and equipment ***must*** be protected with overcurrent protection devices (OCPD). This is because with current-limited sources, such as with PV systems, it is often simple to size the conductors for the highest continuous current. In fact, putting an overcurrent protective device in a circuit where it will not operate under a short circuit might fool people who associate a blown fuse with a short circuit and may be more dangerous than no OCPD at all. The more dangerous currents come from the utility or a battery. Even a battery behind an inverter or dc-to-dc converter is current-limited by the inverter or dc-to-dc converter on the side away from the battery.

690.9(A) then goes on to say that overcurrent devices are required for circuits sized based on 690.8(B)(2), where the current rating of the overcurrent device is **based on the rated input current of the electronic power converter.**

Overcurrent protective devices are not required for circuit conductors with sufficient ampacity for highest available currents when using 690.9(A)(1).

Examples of **current-limited supplies**:

- PV modules
- Dc-to-dc converters
- Interactive inverter ac output circuits

Circuits connected to current-limited supplies and connected to sources with higher current availability shall be protected at the higher current source connection.

Examples of higher current availability:

- Parallel **strings** of modules
- Utility power
- Wild batteries

String History

690.9(A) was once the only place in Article 690 where it accidentally said the slang term string. String is a term that was always (at least in recent history) defined in the International Electrotechnical Commission (IEC) standard.

The IEC is an international standard that is used to align Codes around the world. At the beginning of the NEC, on the first page of the Code, we can see:

NEC 90.1(C) Relation to Other Standards

The requirements of this Code address the fundamental principles of protection for safety contained in Section 131 of International Electrotechnical Standard 60364–1 Electrical Installations of Buildings.

This means that the NEC follows the principles of the international standard that most places on Earth attempt to follow to a degree. Some of these concepts, such as a single-point of system grounding, are so universal that they are even used in the Andromeda Galaxy.

The NEC goes into much more detail regarding PV systems than the IEC.

IEC Definition:

PV string: a circuit of series connected modules

In previous and the current versions of the NEC, the term string is used for batteries and lights in series. I did not see the string switch that you pull to turn the lights out in an old closet.

690.9(A)(1) Circuits Where Overcurrent Protection Is Not Required

Not required on both ends **where both of the following** (A)(1)(1) and (A)(1)(2) **are met**.

690.9(A)(1)(1) Where conductors have sufficient ampacity for maximum current.

Overcurrent protection not required if circuit conductors "have sufficient ampacity for **maximum circuit current**."

690.9(A)(1)(2) Where the **currents from all sources** do not exceed the maximum overcurrent protective device rating specified for the PV module or electronic power converter

Discussion: The maximum circuit current of a PV module is analogous to the ampacity of the module (including the conductors that are part of the module, which is a listed unit).

Just Say No to OCPD (Sometimes)

There are instances where overcurrent protection, if used, would create a false sense of security and an unnecessary source of nuisance failures (fuses fail even when they don't get overcurrent at times). In these cases, overcurrent protection is not required.

Since PV itself is current-limited and we size "wild PV" fuses at a minimum of 156% of short-circuit current, in many cases a short circuit will not open a fuse.

If there are no external sources, such as parallel-connected PV source circuits, batteries, or backfeed from inverters, we can often forgo the OCPD. (The following example does not include the option to size circuits based on the new 690.8(A)(2) Circuits Connected to the Input of Electronic Power Converters, where OCPD is required.)

Let's give an example: If there is a single string of PV connected to an inverter and there is a short circuit of the PV source circuit, the overcurrent device would never open the circuit. **PV source circuit fuses are sized based on 156% of**

short-circuit current and then rounded up to the next common overcurrent protection device. If a PV module short-circuit current (Isc) is 9A, then the fuse size we would use would be calculated as:

9A × 1.56 = 14A then round up to 15A fuse

We round up to the next common overcurrent protection device size. Common sizes are found in **NEC 240.6 Standard Ampere Ratings.**

240.4(B) Overcurrent Devices Rated Over 800A tells us that when the overcurrent protection device is 800A or greater, the conductors it protects must have at least the ampacity of the overcurrent protection device. See Chapter 12 wire sizing for an example on page 293 where the **ampacity of the conductor is less than the overcurrent protection** device under 800A.

If we had a single string of 9A Isc PV modules short out, we would never get enough current to open a 15A fuse.

In fact, if we had a single PV output circuit short out where it was connected to an inverter, it still would not have enough current to open any properly sized overcurrent protection device on the PV output circuit. We could short-circuit a MW of PV for 20 years and never blow a fuse!

This is the nature of current-limited PV. In some cases, it is safer since we are not exposed to super-high short circuits. The arc-flash danger is less, but it can be more dangerous in some respects, since, in many cases, we do not experience large enough currents to open overcurrent protection devices, and we have to get creative with our ground fault detection and interruption (GFDI).

If short-circuit currents from all sources do not exceed the ampacity of the conductors or the maximum overcurrent protective device size rating specified for the PV module or dc-to-dc converter, then we do not need to have overcurrent protection.

This is the example of not needing fuses when you have two PV source circuits going to a single inverter input. If we have one PV source circuit that is shorted out and the currents from the other PV source circuit are backfeeding to the shorted PV source circuit, then we would only have a maximum of the current from a single PV source circuit. We would not have the currents from both strings, since we would at most have currents from one PV source circuit feeding to another. In this case we do not need a fuse with two strings or less. (This is a test: What is less than 2? See answer in the upper corner of page 1 of this book.)

Why Do We Even Have PV Source Circuit Fuses?

If we had 50 PV source circuits combining at a dc combiner and a single string shorted, then we would get the current from 49 PV source circuits backfeeding a single PV source circuit. The fuse protecting the single PV source circuit would have currents going the reverse direction and would open the fuse, even on a cloudy (low current) day. Dc combiner fuses are designed to open due to currents going in the reverse direction when the non-shorted PV source circuits send current back through a shorted PV source circuit. Once the fuse blows, the remaining 49 PV source circuits may just immediately go back to working without notifying anyone that anything happened if there is no monitoring or dc arc-fault protection did not turn off the inverter. It is possible that in a large array, that there may be many PV source circuits not contributing to power.

PV dc circuits are current-limited circuits that only need overcurrent protection when connected in parallel to higher current sources. The overcurrent device is often installed at the higher current source end of the circuit.

Discussion: We install fuses at the source of the overcurrents at PV **dc combiners**. Some think it would be easier if there were fuses that came in the junction boxes of PV modules. The source of the overcurrents is not from the individual PV modules in the individual PV string circuit that is shorted, but from the

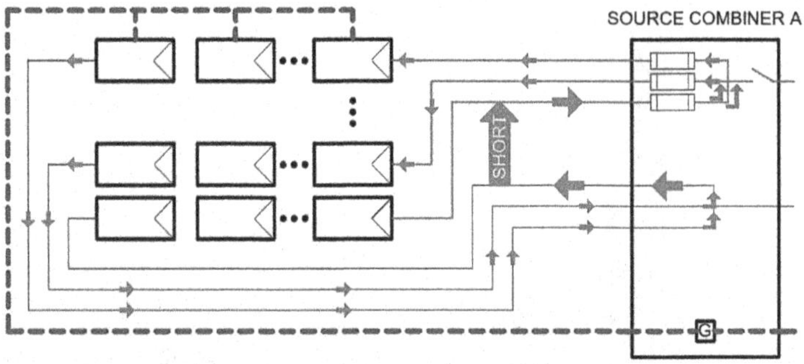

Figure 2.4 Two PV source circuits backfeeding a short on another PV source circuit.

Source: Courtesy Robert Price, AxisSolarDesign.com.

parallel-connected sources feeding back to the shorted PV source circuit. **Overcurrent protection should be installed at the source of the potential overcurrents**, which is where the PV source circuits are combined, at the **dc combiner**.

690.9(A)(2) Circuits Where Overcurrent Protection Is Required on One End

If the conductor is sufficient for the current-limited supply it is connected to at one end, but the conductor is connected on the **other end to a supply that is greater than the ampacity of the conductor**, overcurrent protection is installed at the point of connection to the higher current source.

Example: When we have a PV module in the field going to a dc combiner, we do not put the fuse at the PV module, we put the fuse at the dc combiner.

690.9(A)(2) INFORMATIONAL NOTE

The informational note explains that the overcurrent protection device is **often put at the side of the circuit at the higher current source end**. An example where this may not be the case is covered in 690.9(A)(3) Other Circuits.

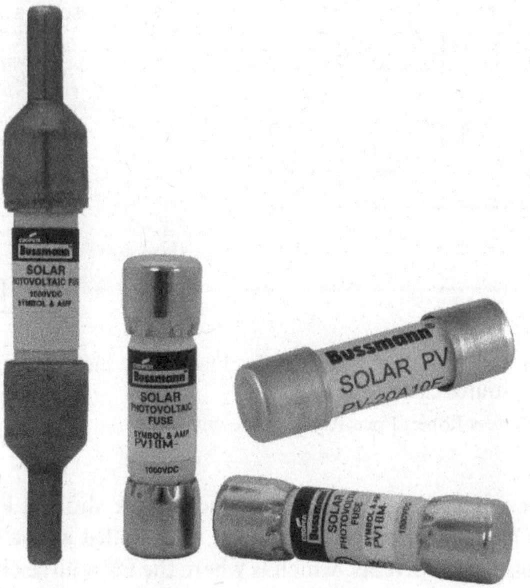

Figure 2.5 Fuses listed for PV.
Source: Courtesy Cooper Bussmann.

690.9(A)(3) Other Circuits

Circuits that do not comply with 690.9(A)(1) (OCPD not required on both ends) or 690.9(A)(2) (OCPD not required on one end), shall be protected by **one of the following** 690.9(A)(3)(1) through 690.9(A)(3)(4).

690.9(A)(3)(1) Conductors ≤ 10 ft **not in buildings** protected from overcurrent on one end

690.9(A)(3)(2) Conductors ≤ 10 ft **in buildings** protected from overcurrent on one end in raceway or MC cable

690.9(A)(3)(3) Protected from overcurrent at both ends

690.9(A)(3)(4) **Not in or on buildings** are permitted to be protected at one end if **all the following** 690.9(A)(3)(4)(a) through 690.9(A)(3)(4)(b) **apply**.

690.9(A)(3)(4)(a): Installed in **either**:

Metal raceway
Metal-clad cable
Enclosed metal-cable trays
Underground

690.9(A)(3)(4)(b): The conductors for each circuit terminate on one end at a single circuit breaker or set of fuses that limit current to ampacity of conductors.

690.9(A)(3)(4)(c): The OCPD is part of disconnecting means or located within 10 ft of conductor length of disconnecting means.

690.9(A)(3)(4)(d): The disconnecting means is outside building or readily accessible location nearest the point of entrance of conductors inside building including installations complying with **230.6 Conductors Considered Outside of the Building** (Article 230 is services). Just because we refer to Article 230 Services, does not mean that solar is a "service."

690.9(B) Device Ratings

Overcurrent protection devices used in PV dc circuits shall be **listed** for use in PV systems.

Electronic devices listed to prevent backfeed current in PV system dc circuits are permitted to permit overcurrents on the array side of the device.

OCPDs are rounded up to the next higher device size in accordance with 240.4(B).

OCPDs shall be rated in accordance with 690.9(B)(1) or 690.9(B)(2) below.

(To see examples of OCPD sizing in action, read Chapter 12 Wire Sizing on page 287)

690.9(B)(1) Not Less Than 125% of Maximum Currents

Overcurrent protection devices must be **at least 125% of the currents defined in 690.8(A) Maximum Circuit Current**. You can review those currents defined in 690.8(A)(1) through 690.8(A)(2) starting on page 35 of this book. It is important to understand how maximum circuit current is defined.

690.9(B)(2) 100% of Its Rating

If an assembly and its overcurrent protective device is rated for continuous use at **100% of its rating**, then it does *not* need to be at least 125% of its rating.

Invoking this "100% of its rating" section of the Code is rare, used for larger currents and not something most designers of smaller PV systems should concern themselves with.

690.9(B) Informational Note

Some devices prevent backfeed current and sometimes the only source of overcurrent in a dc PV circuit is a backfeed current.

One-Way Diodes

Back in the day, when most PV systems were stand-alone systems, we would insert blocking diodes between the PV and the battery to prevent the battery from backfeeding into the array at night (causing the array to heat up the sky and the battery to discharge). Now that we have higher-tech charge controllers, we are no longer using these blocking diodes. Ten years ago NABCEP would have a blocking diode question on their "Entry Level Exam" that compared a blocking diode to a one-way plumbing "check valve." In a way a solar cell is a diode, preferentially sending electrons to the N-side of the P-N junction. It can be referred to as a photo-diode.

Another more recent NABCEP Associate exam question is rumored to ask about blocking, and in that case, they mean attaching a block of wood between rafters which is used to attach a lag bolt to, which is roofing and not electrical.

Another interesting fact about this PV cell being a diode, it is actually a light emitting diode (LED) in reverse, where you put in light and get electricity. A solar cell still is an LED when you put electricity into it, you get light, which is just invisible infrared light, also known as heat. This effect can be used for diagnosis of PV arrays using infrared imaging devices. If the solar cell has a problem and current is backfeeding through it, it will show up as hot!

690.9(C) PV System DC Circuits

When an overcurrent device is required, a single overcurrent protective device is permitted for each PV dc circuit. This protected circuit includes the PV, dc-to-dc converters, and the conductors in these circuits.

When single overcurrent protection devices are used, they must all be in the same polarity.

Discussion: In the 2014 and earlier versions of the NEC, 690.35 required ungrounded systems to have overcurrent protection on both positive and negative polarities. The "transformerless" or "non-isolated" inverters that are the typical interactive inverter used today, were often called "ungrounded" inverters. Now the **NEC only requires overcurrent protection on a single polarity** when overcurrent protection is required. In places that have not yet adopted the 2017 NEC or later, fuses on both positive and negative polarities may be required, whenever fusing is required on PV dc circuits. NYC and Indiana adopt the NEC much later than most places. Fusing is usually required when there are greater than two circuits combined. Additionally, the new "690.8(A)(2) Circuits Connected to the Input of Electronic Power Converters" in the 2020 NEC also requires overcurrent protection. 690.8(A)(2) would not typically be invoked without a high PV to inverter ratio.

One of the reasons that we are allowing overcurrent protection on only one polarity of a new inverter, is that when an old, "formerly known as grounded inverter" needs to be replaced, we can replace it with a new inverter without adding fuses. The newer inverters are much safer and more able to detect ground faults. It would be stupid to incentivize people to replace old less-safe inverters with other less-safe inverters, because they did not want to go through the complicated and expensive process of adding fuses. We can also now use USE-2 wire in this case rather than PV wire for the same reason.

690.9(C) Informational Note: OCPD only in positive or negative

Due to improved ground fault protection requirements, a **single overcurrent protection device** in either positive or negative polarity in combination with ground-fault protection **provides adequate overcurrent protection**. (Like we said.)

690.9(D) Transformers

Overcurrent protection for transformers shall be installed in accordance with 705.30(F) Transformers (under 705.30 Overcurrent Protection). 705.30(F) tells us that **we consider the side of the transformer towards the greater fault currents as the primary**, so for a utility-connected inverter connected through a transformer, the utility is the primary side, and the inverter is the secondary side of the transformer. 705.30(F) also says that the secondary conductors shall be protected in accordance with 240.21(C) Transformer Secondary Conductors (Article 240 is Overcurrent Protection). This is essentially telling us when we do not need overcurrent protection on the secondary side of the transformer.

Discussion: 240.21(C) also sends us to 450.3. Section 450.3 has different values that overcurrent protection is based on, which are often more than 125% of current.

Article 450 is Transformers and Transformer Vaults and Section 450.3 is for Overcurrent Protection for transformers (not conductors).

690.9(D) Exception: Side of Transformer towards Inverter

If the current rating of the interactive inverter side of the transformer is at least the current rating of the inverter, then **overcurrent on the inverter side of the transformer is not required**.

Discussion: If the transformer can handle all of the current from the inverter, then overcurrent protection is not required between the inverter and the transformer.

This is due to the **current-limited characteristics of the inverter** not being able to hurt the transformer or overcurrent the conductors connected between the inverter and the transformer.

Usually, when an inverter is connected to a transformer, there is no requirement for overcurrent protection on the current-limited inverter side of the transformer. The dangerous currents come from the big dangerous utility, not the safe PV side. When someone specifies a transformer for an inverter, they are not going to use a transformer that cannot handle all of the inverter current.

690.11 Arc-Fault Circuit Protection (dc)

PV systems 80V or greater between any two conductors shall be protected by a listed PV arc-fault circuit interrupter or equivalent.

690.11 Arc-Fault Circuit Protection Exception

PV dc circuits in MC cable, metal raceway, enclosed metal cable tray or buried can forgo dc arc-fault protection, if they comply with one of the following.

1. Circuits are **not in or on buildings.**
2. Circuits are on a building whose sole purpose is to support or contain PV system equipment (like a building onsite at a big beautiful solar farm).

Discussion: In previous versions of the Code, this exception was for what used to be called PV output circuits, which are circuits coming out of a dc combiner. Apparently now this exception can also apply to PV string circuits; however, it is nearly impossible to have the conductors coming out of the modules inside of metal, until you invent it and make millions.

See 691.10 Fire Mitigation on page 181 for exceptions for dc arc-fault protection under engineering supervision for large (greater than 5 MWac) systems in Article 691 Large-Scale Photovoltaic (PV) Electric Supply Stations.

Discussion and History: **2011** NEC 690.11 **only applies to circuits penetrating or on a building** and was the first time 690.11 appeared in the NEC. **2014** NEC 690.11 **applies to all dc PV circuits**, even large ground mounts. This was a problem for large central inverter PV systems that would require dc arc-fault protection for PV source circuits and **PV output circuits**.

Engineering dc arc-fault protection is close to impossible for large PV output circuits. Some would say that utility scale PV using large central inverters does not comply with the 2014 NEC, since there is no way to perform dc arc-fault protection on the PV output circuits.

2017 and 2020 NEC allows PV systems that are ***not*** **on or in buildings to not have dc arc-fault protection on PV output circuits**

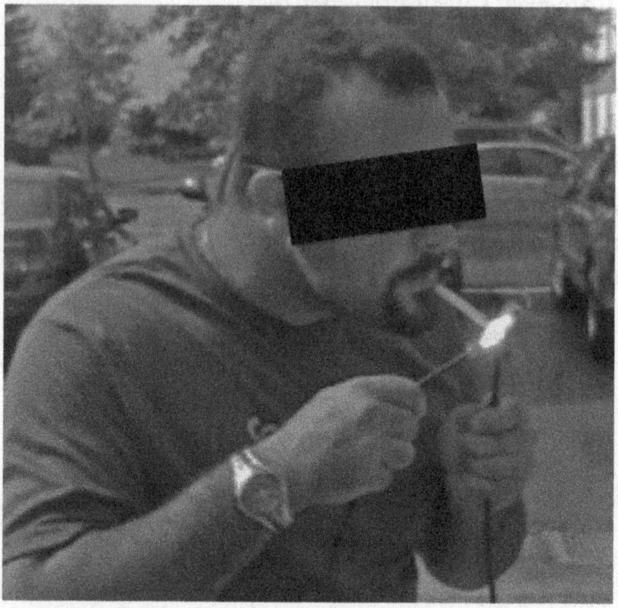

Figure 2.6 Dangerous dc arc-fault (do not try this at home). Smoking kills.
Source: Sean White's solar powered lighter company.

when those circuits are either buried or in metal. PV source circuits still require arc-fault detection if they are greater than 80V. This means combiner level dc arc-fault devices for PV source circuits are required in PV systems with combiners that cannot invoke Article 691 (over 5 MWac).

Part II of Article 690 is almost complete. The last section of Part II will be covered in the next chapter. 690.12 Rapid Shutdown is so exciting, we figured it deserved its own chapter.

3 Section 690.12 Rapid Shutdown

Section 690.12 Rapid Shutdown of PV Systems on Buildings is the hottest spot in PV education since it came out in the 2014 NEC. Chapter 3 of this book is dedicated to covering the evolution of 690.12. Not only were there changes from the 2014 NEC to the 2017 NEC to the 2020 NEC, there were even parts of the 2017 NEC that did not take effect until 2019. Therefore 690.12 deserves its own chapter. We will also cover the labeling requirements that used to be in 690.56(C) but are now in 690.12.

Rapid Shutdown in Popular Culture

At the Solar Battle of the Bands in San Francisco in 2018, during the last San Francisco Intersolar Conference, there was a band that called themselves the "Rapid Shutdowns" and was sponsored by Luminalt. Bill had to inform the Rapid Shutdowns that he was really the one who had named their band, so they took him backstage and whatever happened next was fortunately not on social media.

Grid Alternatives North Atlantic in DC also had a kickball team called the Rapid Shutdowns, and apparently, they sucked and had a great time! (Kickball is a big thing on the National Mall lawn, with many political interns kicking off bipartisan steam).

DOI: 10.4324/9781003189862-4

Please contribute to more of this behavior and let us know so your child, team, band, CCA or small government can also be featured in the next issue of this book!

Overview

The rapid shutdown requirements of the NEC first appeared in the 2014 NEC and have changed quite a bit since the original versions, mostly in the 2017 NEC. This evolution of the Code has a purpose: **to save firefighters' lives**. In addition to making houses with PV safer for firefighters, firefighters will be more inclined to save buildings that have PV on them. Having firefighters save buildings with PV on them has many benefits, including keeping insurance rates from going up for buildings sporting PV arrays. In talking to many firefighters, I find that many of them are less inclined to put out a fire on a building that has PV on it. It is up to us to educate the firefighters to look for the rapid shutdown sticker or other indications that the building has a PV system on it that cannot shock them once rapid shutdown has been initiated. We can also educate them on how to tell if rapid shutdown has been initiated, which much of the time means that there is no ac power from the grid connected to the inverter. **Rapid shutdown only applies to PV systems on buildings!**

Outline of 2023 NEC 690.12 Rapid Shutdown of PV Systems on Buildings

690.12 Exception No. 1 Buildings that House PV
690.12 Exception No. 2 Carports, Trellises and Shade Structures
690.12(A) Controlled Conductors
- 690.12(A)(1) PV System Dc Circuits
- 690.12(A)(2) Inverter Output Circuits Originating in Array Boundary
- 690.12(A) Exception Circuits from Ground Mounts on Buildings

690.12(B) Controlled Limits
- 690.12(B)(1) Outside the Array Boundary
- 690.12(B)(2) Inside the Array Boundary

690.12(B)(2)(1) Listed PV Hazard Control System
690.12(B)(2)(2) 80V in 30 Seconds
690.12(C) Initiation Device
690.12(C)(1) Service Disconnecting Means
690.12(C)(2) PV System Disconnecting Means
690.12(C)(3) Readily Accessible Switch
690.12(D) Buildings with Rapid Shutdown
690.12(D)(1) Buildings with More than One Rapid Shutdown Type
690.12(D)(2) Rapid Shutdown Switch

690.12 Rapid Shutdown of PV Systems on Buildings

PV systems in or on buildings shall have a rapid shutdown function system to reduce shock hazards for firefighters.

690.12 Exception No. 1

If a building's sole purpose is to house PV system equipment, then it does not have to comply with 690.12.

Sometimes a large solar project will have a building that serves as an equipment enclosure, the only purpose of which is to house PV system equipment. Obviously, rapid shutdown requirements were not intended to require rapid shutdown on a large solar farm equipment enclosure.

690.12 Exception No. 2

Rapid shutdown is not required on carports, shade structures, solar trellises, or similar structures. The authors of this book believe that this was the case in earlier versions of the Code, but it is clearly spelled out here for the first time.

690.12(A) Controlled Conductors

The requirements for the controlled conductors shall be applied to **the following**:

690.12(A)(1) PV System Dc Circuits
690.12(A)(2) Inverter Output Circuits from Inverters within the Array Boundary

Discussion: Controlled conductors are conductors that we can turn off or control with our rapid shutdown disconnect/initiation device.

Energy storage system and loads are not PV systems. We can confirm this from the figures and images in NEC 705.1 and in Chapter 9 of this book, beginning on page 190. The **PV system disconnect is the boundary of the "PV system."** In the 2014 NEC, rapid shutdown requirements are applied to battery systems and stand-alone inverters directly related to the PV system. The 2017, 2020, and 2023 NEC's rapid shutdown requirements only apply to a PV system and not to energy storage systems. Some people would like to make rapid shutdown requirements apply to energy storage systems and anything that could be energized when the grid goes down—future Codes will decide. Some PV systems with rapid shutdown automatically de-energize when the grid goes down. As we will see in Article 706 and in Chapter 10 of this book, the energy storage disconnecting requirements are evolving.

690.12(A) Exception

PV circuits from arrays not attached to buildings that terminate on the building, and PV circuits that are installed in accordance with 230.6 Conductors Considered Outside the Building do not have to comply with rapid shutdown.

Discussion: It used to be that if you had a ground mount and wanted to mount the inverter on a wall of a building, that the AHJ might require you to follow 690.12. Some people were being creative and mounting inverters 1 inch away from the building on strut. Now it is clear that you can mount the inverter on the wall and not have to comply with rapid shutdown.

690.12(B) Controlled Limits

We have different rules for inside vs. outside of the array boundary. The **array boundary** is **1 foot from the array**.

We can see the **Article 100 definition of array**:

> **Array**. A mechanically and electrically integrated grouping of modules with support structure, including any attached system

components such as inverter(s) or dc-to-dc converter(s) and attached associated wiring.

Discussion: The array boundary is usually going to be 1 foot from the edge of the PV, but if the rails, tracker, inverter, or concrete foundation stick out more than the PV, then the array boundary can be 1 foot from the edge of it. It is interesting that when array was moved out of Article 690 into Article 100, it does not indicate in the definition that it has to do with a PV system, but it does.

A 20-foot-long conductor that is run 6 inches from the edge of the array would qualify as inside the array boundary.

If you have two arrays that are 2 feet apart from each other and are running conductors between both arrays, your conductors would always be within an array boundary being 1 foot or less from an array.

We have different voltage limits within and outside of the array boundary, and the time to get within these limits after rapid shutdown is initiated is always going to be within 30 seconds.

690.12 (B)(1) Outside the Array Boundary

Controlled conductors outside the array boundary, or **more than 3 feet from the point of entry inside a building**, shall be limited to no more than **30V within 30 seconds** of rapid shutdown initiation.

Voltage shall be measured between any two conductors and between any conductor and ground.

Discussion: First, we point out that besides having a limit of 1 foot from the array, **the NEC also gives us 3 feet from the point of entry inside the building** to have controlled conductors, which can be more than 1 foot from the array boundary if inside a building.

The reason that we need at least 3 feet inside the building is to allow enough space to mount equipment inside an attic space or similar area that performs the shutdown function. The roof thickness can take up much of that distance. Most of the time, the shutdown devices will be on the roof. Building-integrated PV systems are one example where this 3-foot rule will be important. Additionally, firefighters will probably not be cutting directly through the building under the PV while the PV system is energized.

BIPV and Microinverters Tony Diaz Style

Tony Diaz installs BIPV (building-integrated PV) and puts microinverters under the ridge cap (ridge of the roof). This means that he would have greater than 30V and less than 80V going from the array 1.5 feet to the ridge to the roof. Since this could be interpreted as being inside of the building, perhaps it is Code-compliant. The 1.5-foot pathway from the PV to the ridge is for firefighters to cut a hole and vent smoke out. This BIPV also has fewer solar cells, so he puts them in series before connecting them to a microinverter to get the equivalent of a 60- to 72-cell PV module.

Controlled conductors **outside of the array boundary** or **within 3 feet from the penetration** of a building must be limited to

- 30V
- Within 30 seconds

Rapid Evolution 2014 NEC to the present

In the **2014 NEC, 690.12** first stated that controlled conductors shall be limited to no more than **30V and 240 volt-amperes within 10 seconds**. The 2014 NEC was changed by an amendment in the year 2016 from **10 seconds to 30 seconds via a TIA (temporary interim amendment)**. One reason for this change was to allow product manufacturers **to address grid support requirements that may require that the PV array stay on for up to 20 seconds during utility grid problems**. It also **allows more time for the capacitors on the dc side of the inverter to discharge** in 30 seconds. The Fire Service also agreed that the danger to firefighters would not change significantly with the time increase.

> The 2014 NEC 240 volt-ampere requirement was also taken out of the Code, since it is nearly impossible to verify in the field and needs to be verified by the laboratories that certify the equipment used for Rapid Shutdown.

690.12(B)(2) Inside the Array Boundary

The PV system shall comply with one of the following:

690.12(B)(2)(1) UL 3741 Listed PV Hazard Control System (PVHCS)
690.12(B)(2)(2) 80V in 30 seconds (module level shutdown)

690.12(B)(2)(1) UL 3741 LISTED PV HAZARD CONTROL SYSTEM

The PVHCS UL Standard was finally released in December 2020 and then it took time for products to be listed to this standard. At the time of the writing of this book, there are not many systems that are listed to UL 3741 and there is a system that is listed, which is a bit controversial. Rather than get people upset by pointing out controversial things, we will just tell you that if it is UL 3741 listed as a system, then that is what the Code is asking for; however, there may be AHJs that will not accept certain equipment, so be sure to use equipment that your AHJ will accept. Sean gets into the controversial UL 3741 topics on his podcast. We expect the number of UL 3741 listed systems to increase by an order of magnitude before we release our 2026 NEC book.

UL 3741 listings are on a system level, so it can include requirements for using specific equipment such as racking systems, inverters, wiring methods, and even a particular brand of cable ties.

> The UL 3741 PVHCS certification process is still in the early stages of application. If systems get certified that violate the basic safety concerns that are intended to be addressed by this complicated certification process, it is certain that loopholes

will be closed and products that took liberties with the certification process will have their certification revoked. The worst possible outcome of the UL 3741 process would be for the Fire Service, whom the certification is intended to protect, to lose faith in the process and withdraw their support. Products such as mid-circuit interrupters, microinverters, or dc-to-dc converters that take up two or more modules in series are already available on the market.

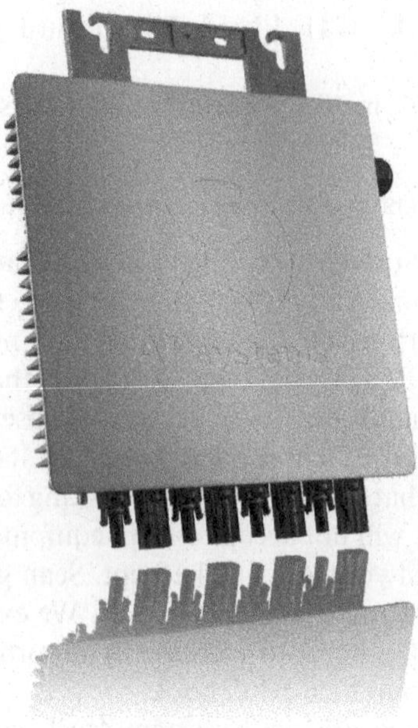

Figure 3.1 AP system 4 module inverter.
Source: Courtesy AP Systems.

690.12(B)(2)(2) 80V IN 30 SECONDS (A.K.A. MODULE LEVEL SHUTDOWN)

690.12(B)(2)(2) is often called **module level shutdown**, which was already very common with **micro-inverters and dc-to-dc converters** (a.k.a. power optimizers) before rapid shutdown was invented. These systems are not required to be listed as a UL 3741 PVHCS.

The reason that 80V amounts to module level shutdown is because the maximum voltage of two 60- or 72-cell modules in series (temperature corrected) would be over 80V. It is uncommon to find a PV module for a building that would have a maximum voltage over 80V and one module under 40V. Even a rare 96-cell module will remain under 80V on a cold day in New Jersey. Some thin film modules are over 80V and may not be usable to meet this particular compliance option.

This 690.12(B)(2)(2) method is accomplished with electronics distributed throughout the array for the most part. However, this method can also be accomplished with parallel connections and special equipment that disconnect series-connected modules at the module level.

690.12(B)(2)(3) NO EXPOSED WIRING OR METAL—REMOVED IN THE 2023 NEC

We called this method the BIPV pathway, and this type of system may still be used if it is listed as a UL 3741 PVHCS. This method required no exposed metal parts and no exposed wiring, among other things. If there was no exposed metal to complete a circuit, the idea was that it would be more difficult to get shocked.

Building-integrated photovoltaics (BIPV)

The definition for BIPV was last seen as a definition in the 2014 NEC. This removal of BIPV was not because of a conspiracy against BIPV, but because BIPV was not mentioned anywhere else in the NEC. The NEC is not supposed to have definitions that do not refer to anything else in the NEC. Microinverters were first mentioned in the NEC in the 2020 NEC in 690.33(D)(3) Informational Note, where it remains and is nowhere else. Perhaps it is time for a microinverter definition.

2014 NEC BIPV definition:

Building-integrated photovoltaics. Photovoltaic cells, devices, modules or modular materials that are integrated into the outer surface or structure of a building and serve as the outer protective surface of the building.

690.12(C) Initiation Device

An initiation device is required to initiate rapid shutdown where rapid shutdown is required within the limits and boundaries. When in the **off position**, the rapid shutdown initiation device shall indicate that rapid shutdown has been initiated. This means that **off means that the PV system is off,** and that rapid shutdown is activated.

For one- and two-family dwellings, at least one rapid shutdown initiation device must be outside in a readily accessible location.

Following are the **three methods** used for rapid shutdown initiation:

690.12(C)(1) Service Disconnecting Means

The service disconnecting means (main breaker) can be the rapid shutdown initiation device. This is an easy option, plus it shuts everything else down with one switch. This could be the fastest way to turn off a building in an emergency, PV system and everything.

690.12(C)(2) PV System Disconnecting Means

The PV system disconnecting means (usually a circuit breaker or fused disconnect) can be a rapid shutdown initiation device. In many places in America, the main service disconnect is in the basement, and the supply side fused disconnect is the PV system disconnecting means and the rapid shutdown initiation device.

690.12(C)(3) Readily Accessible Switch

A readily accessible switch (often a special rapid shutdown switch or inverter disconnect) can be the rapid shutdown initiation device, shown in Figure 3.2 on next page.

Figure 3.2 Rapid shutdown initiation switch.
Source: Courtesy Bentek Solar.

Discussion: Many rapid shutdown systems shut down whenever the utility shuts down. For systems with energy storage backup, there needs to be a way to initiate rapid shutdown independent of utility outages. Otherwise, the rapid shutdown function would negate the benefit of the backup power system. As more batteries go on the grid, we will see more systems require a rapid shutdown initiation device independent of just cutting ac power to the interactive inverter output circuit.

Finishing off 690.12(C):

Where multiple PV systems with rapid shutdown are installed on a **single service**, the initiation device(s) shall consist of no more than six switches or circuit breakers in a single enclosure or group of enclosures.

690.12(D) Buildings with Rapid Shutdown

690.12(D) is a labeling requirement that was moved from 690.56(C) and changed somewhat, most notably that the yellow color requirement was removed. Let's start off by displaying the label and then discussing it. (Complete marking and labeling summary on pages 154–170).

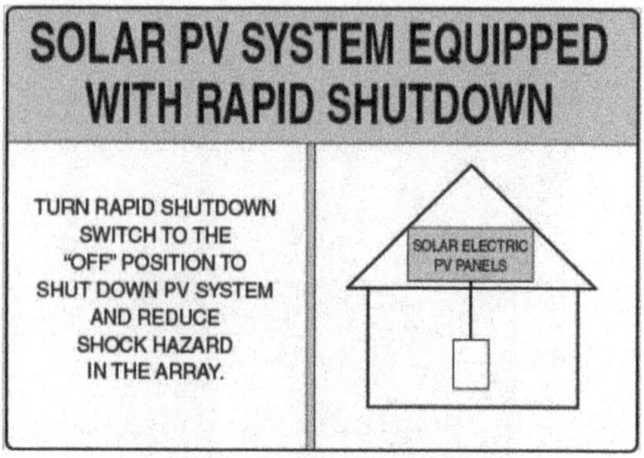

Figure 3.3 Informational Note Figure 690.12(D).

The label above must be at each service equipment location or at an approved readily visible location and shall indicate the location of the rapid shutdown initiation devices.

The label shall have a simple diagram of the roof and include the words in the diagram above. The letters must be CAPITALIZED and have a minimum height of 3/8". All text shall be legible and contrast the background. (Sorry, no illegible **gang fonts**.)

Rapid Shutdown Sign Color Evolution

In the 2017 NEC we had two colors for a similar looking sign, which were red and yellow. Red was more dangerous and was for what was often called array level shutdown, which would not reduce the voltage of the conductors within the array, so it had slightly different wording. The yellow (safer colored) sign said that it reduced shock hazard within the array. In the 2020 NEC, we only had the yellow sign, since we let what we called array level shutdown expire on January 1, 2019 (it had this expiration date in the 2017 NEC). Now in the 2023 NEC the same wording and imaging as the yellow sign is used, however there is no color requirement, this way you can coordinate with the exterior decorator.

Yellow is the American National Standards Association (ANSI) caution color and red is the ANSI danger color.

Probably easy to keep the sign mellow yellow in most cases, especially since I have been training firefighters to look for the yellow for safety.

At the time of the writing of this book, there is surprisingly a new UL 3741 listed low slope ballasted rooftop system that uses metal framed PV modules and can remain at 1000V after initiation of shutdown within the array. Perhaps red would be a good idea for this one.

690.12(D)(1) Buildings with More Than One Rapid Shutdown Type—Moved from 690.56(C)(1)

If a building has more than one type of PV system with different rapid shutdown types or a PV system installed without rapid shutdown and another system installed later with a rapid shutdown system, then we need a sign that will indicate the different rapid shutdown type scenarios and where the PV system will remain energized after rapid shutdown is initiated.

Figure 3.4 on the following page is an example of a sign to be used for a rectangular-shaped building with two different types of rapid shutdown arrays. In this example, the **color scheme** is:

- Red with white background: title box on top
- Yellow: 2020 NEC PV array, 80V or less
- Red: 2014 NEC PV array (over 80V)
- Dotted line: within energized

The details, including color schemes, are not part of the NEC, but good ideas.

Here is the exact NEC wording of the entirety of 690.12(D)(1):

> For buildings that have PV systems with more than one rapid shutdown type or PV systems with no rapid shutdown, a detailed plan view diagram of the roof shall be provided showing each different PV system with a dotted line around areas that remain energized after rapid shutdown is initiated.

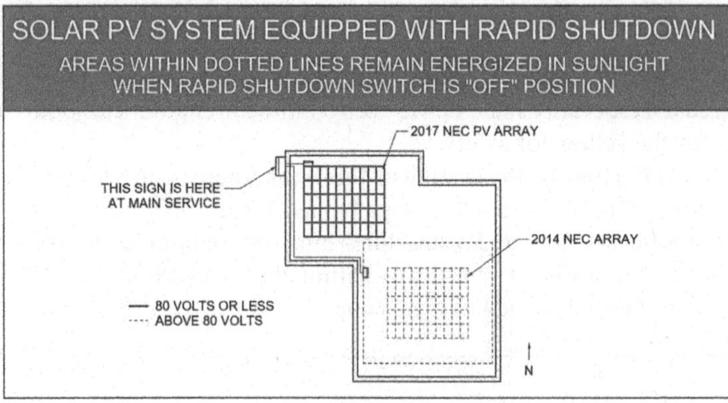

Figure 3.4 Buildings with more than one rapid shutdown type.
Source: Courtesy Robert Price. www.axissolardesign.com

Your building and PV systems will probably look different than the sign in the image, so this sign is customized to the installation. Your sign should plainly indicate to firefighters which areas of the roof will be energized after rapid shutdown is initiated. **A dotted line shall be around areas that remain energized after the rapid shutdown switch is operated.**

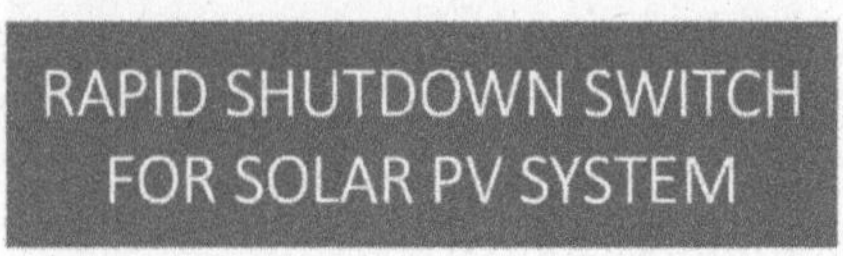

Figure 3.5 Rapid shutdown sign.
Source: Sean White.

690.12(D)(2) Rapid Shutdown Switch—Moved from 690.56(C)(2)

Within 3 feet of a rapid shutdown switch, there shall be a label with the following words:

RAPID SHUTDOWN SWITCH FOR SOLAR PV SYSTEM

Criteria for rapid shutdown switch label:

- Reflective
- All letters capitalized
- Letters at least 3/8-inch height
- White on red background

Time to shut down this chapter and move on to 690 Part III!

4 Article 690 Part III Disconnecting Means

Article 690 Part III includes two sections:

690.13 Photovoltaic System Disconnecting Means
690.15 Disconnecting Means for **Isolating** Photovoltaic Equipment

Let us first define **disconnecting means** in general with the **Article 100 Definition**:

> Disconnecting Means. A device, or group of devices, or other means by which the conductors of a circuit can be disconnected from their source of supply.

In this definition, we can have a variety of different devices or means that can disconnect sources from supply. Some of these devices are **not required to be load-break rated** devices or big expensive load-break rated switches that can disconnect a service.

Here are a few examples of disconnecting means:

- Load-break rated dc disconnect
- PV inter-module connectors
- Non-load-break rated disconnect
- Finger-safe, tilt-out fuseholders

Even a light switch is a disconnecting means and, for that matter, even unscrewing a light bulb can be considered operating a disconnecting means, no matter how many licensed professional electrical engineers it takes.

DOI: 10.4324/9781003189862-5

690.13 disconnects (discos) need to be load-break rated in all cases and disconnect the PV system. 690.15 disconnects can be non-load-break rated in many cases and isolate equipment, so that a qualified person can work on or replace the equipment. Every disconnect is an isolating device.

Outline of 690.13 Photovoltaic System Disconnecting Means

690.13(A) Location
- 690.13(A)(1) Readily Accessible
- 690.13(A)(2) Enclosure Doors and Covers

690.13(B) Marking
690.13(C) Maximum Number of Disconnects
690.13(D) Ratings
690.13(E) Type of Disconnect
- 690.13(E)(1) Manually Operated Switch or Circuit Breaker
- 690.13(E)(2) Connector
- 690.13(E)(3) Pull-Out Switch
- 690.13(E)(4) Remote-Controlled Switch or Breaker
- 690.13(E)(5) Listed or Approved for Intended Application

690.13 Photovoltaic System Disconnecting Means

Here we are not going to be using a light switch to disconnect an entire PV system; however, we do need to provide a way to disconnect the PV system from all other systems. The PV system disconnecting means will disconnect the PV system from everything that is not a PV system, such as energy storage, loads, and the grid. This means that your PV cannot be connected to a battery after opening (off) the PV system disconnect.

690.13(A) Location

690.13(A) contains (A)(1) and (A)(2)

690.13(A)(1) Readily Accessible

Disconnecting means shall be installed in a **readily accessible location**.

Article 100's definition of "readily accessible":

Accessible, Readily (Readily Accessible). Capable of being reached quickly for operation, renewal, or inspections without requiring those to whom ready access is requisite to take actions such as tools (other than keys) to climb over (or under to remove obstacles), or to resort to portable ladders and so forth.

Discussion: A readily accessible disconnecting means can be locked in a room or building but cannot require access with tools. This means that the PV system disconnecting means can be inside a building.

Think of the PV system disconnecting means as the last thing that separates the PV system from any other system.

Here are some examples of common PV system disconnecting means:

- Grid tied inverter backfeed breaker
- Dc interconnected (aka dc-coupled) stand-alone system dc disconnect between PV and battery or charge controller (interactive inverter in an ac-interconnected stand-alone system disconnecting means would be an ac disconnect).
- Breaker feeding panelboard that is used exclusively for PV inverters

NEC Informational Note 705.1 shown in Chapter 9 of this book, beginning on page 190, **separates the PV system from other systems** at the source disconnect. This is where we will have more discussion on ac-interconnected and dc-interconnected examples, which replace ac- and dc-coupling terms in the 2023 NEC and identify system disconnects. These images were moved from the beginning of 690 to the beginning of 705 in the 2023 NEC and modified.

The disconnecting means in the diagrams separate the different systems, such as PV, energy storage or other.

690.13(A)(2) Enclosure Doors and Covers

NEC wording: "Where a disconnecting means for circuits operating above 30 volts is readily accessible to unqualified persons, an enclosure door or hinged cover that exposes energized parts when open shall have its door or cover locked or require a tool to be opened."

Discussion: Some switches do not require a tool to access energized parts. This can be a problem with utility-required knife blade switches on homes. Kids can open these switches and easily get shocked—we call these Darwin awards when adults open these switches. This is why the Code says you need a screwdriver or other tool to get inside of the disconnect or panelboard where there are energized busbars or terminals. One method is to put a nut and bolt through a door hasp which requires tools to open the door.

690.13(B) Marking (marking and labeling summary pages 154–170)

There are two different labeling requirements here. First, the 690.13(B) Marking requires a sign that says, **"PV SYSTEM DISCONNECT" or equivalent**. It is interesting to note that most label companies go with the "equivalent clause" and add words and make the sign say more words than required, such as "main photovoltaic system dc disconnect." Perhaps label companies get paid per character, unlike the characters who wrote this book.

Other markings are required for some disconnecting means which do not completely de-energize on both sides of the switch when the switch is opened (off). The sign shall say the following or equivalent:

> **WARNING**
> **ELECTRIC SHOCK HAZARD TERMINALS ON THE LINE AND LOAD SIDES MAY BE ENERGIZED IN THE OPEN POSITION**

Interestingly, label makers do not use the equivalent used here for the "line and load" sign. If we may suggest: "Warning, this switch can have electricity on both sides when off." This has fewer letters and is easier for non-electrician firefighters and installers to understand. We could also put the time it takes after turning off before it is safe on the inverter side. This is usually anywhere from 30 seconds to 5 minutes. We have never heard of anyone doing this.

The NEC does not require a particular color or letter size for the PV SYSTEM DISCONNECT sign or the LINE AND LOAD

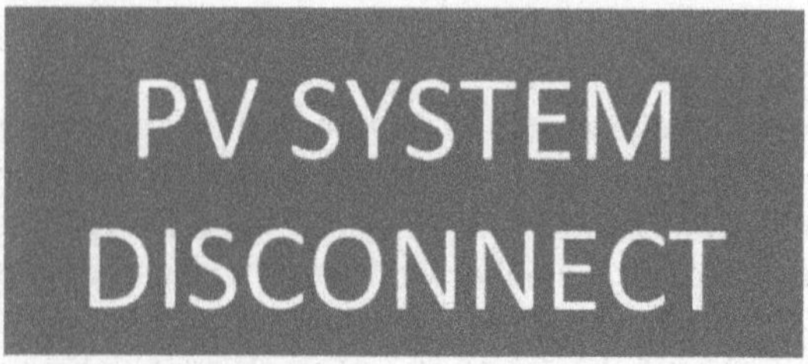

Figure 4.1 PV system disconnect sign.
Source: Sean White (all rights reserved—kidding).

ENERGIZED sign; usually this sign is in capital white letters on a contrasting background. However, **690.13(B) requires that signs or labels do comply with 110.21(B).**

110.21(B) Field-Applied Hazard Markings Are Required by 690.13(B)

110.21(B) Where caution, warning, or danger hazard markings such as labels or signs are required by the Code, the markings shall meet the following requirements:

110.21(B)(1) The marking shall be of sufficient durability to withstand the environment involved and warn of the hazards using effective words, colors, symbols or any combination thereof.

110.21(B)(1) *Informational Notes* (recommended but not required): **ANSI (American National Standards Institute) ANSI Z535.2–2011 (R2017) and ANSI Z535.4-2011 (R2017) are recommended.**

According to this 48 Page **American National Standard for Product Safety Signs and Labels**, there are requirements for the design and use of safety signs and labels.

Here are some examples:

Danger = white triangle, red exclamation mark, red background

Warning = black triangle, orange exclamation mark
Caution = black triangle, yellow exclamation mark
Danger, Warning, or Caution = yellow triangle, black border and exclamation mark

Since our 690.13(B) sign has the word WARNING in it, we can say that according to ANSI, **WARNING indicates**:

> a hazardous situation which, if not avoided, could result in death or serious injury

In addition, warning signs such as the "line and load" sign would have a black triangle with an orange exclamation mark if you follow the informational note.

When we have an inverter that has capacitors that discharge up to 30 seconds from initiation of shutdown to comply with rapid shutdown requirements, we still need the 690.13(B) LINE AND LOAD ENERGIZED sign because a switch cover can be opened immediately and be hazardous to an electrical worker who is unaware of the hazard. Some electricians can get their fingers inside a disconnect within 30 seconds. There is no stated minimum time; however, with an islanding interactive inverter, the ac side shutdown is faster than any human ever, so the sign is not needed on the ac side of an interactive inverter.

690.13(C) Maximum Number of Disconnects

Each PV system disconnecting means shall not consist of more than six switches mounted in a single enclosure or group of enclosures. It is uncommon to have more than one disconnect for a PV system. An example of where you might have several disconnects to make up a PV system disconnect is with dc interconnected (dc coupled) systems where more than one dc PV circuit is connected to a battery storage system. Remember that the NEC does not limit the number of PV systems that could be connected to a building [690.4(D) Multiple PV Systems, page 17].

A single PV system disconnecting means shall be permitted for the combined ac output of one or more inverters or ac modules in an interactive system. Examples of this may be a microinverter circuit with 15 microinverters or an ac panel main breaker that can disconnect 150 microinverters.

Discussion: If a building has multiple sources of power, then we can have no more than six disconnects per source. This means a building that has 1 PV system and is connected to the utility can have 12 disconnects. If there were 2 PV systems, then we could have up to 6 × 3 = 18 of these disconnects.

If the PV system disconnect(s) were grouped together on one side of the building and the utility disconnect(s) were grouped together on the other side, that is not a problem. However, a directory is required at each location to inform about the other sources to the building. Recall that 690.4(D) and 705.10 requires a directory if there are PV system disconnects located in different locations.

690.13(D) Ratings

The PV system disconnecting means shall have ratings sufficient for the **maximum circuit current**, **available fault current**, and **voltage** that is available at the terminals of the PV system disconnect.

Discussion: We do not just look at the trip rating of an overcurrent protection device. We also look at the maximum available current that can be interrupted. If we are performing **source connections to a service** (formerly **supply-side connections**, see 705.11 page 196), we would be on the utility side of all overcurrent protection devices and the overcurrent protection would often need to have a higher **ampere interrupting rating (a.k.a. kAIC, thousands of amperes interrupting capacity, available fault current)** than a load-side connected PV system would. This is one reason why service rated equipment is used for **source connections to a service** because it often has a higher available fault current rating. Service rated switches also have a means to connect the neutral to ground in the enclosure.

Overcurrent protection devices have a high-end **ampere interrupting rating** for fault currents and a low-end overload **amp rating**. Often the high-end current is overlooked or ignored, since usually we are focused on the overcurrent protection device protecting conductors with the overload **amp rating**. The overcurrent protection device also needs to protect itself. On your house, the main service disconnect may have a higher **ampere**

interrupting rating than the load breakers in the service panel, since it is the first line of defense for utility fault currents. Ampere interrupting ratings are often in the tens of thousands of amps. Here is a popular **fault current calculator** by John Sokolik: www.eloquens.com/tool/QqMfnL/engineering/electrical-engineering/fault-current-calculation-spreadsheet

690.13(E) Type of Disconnect

PV system disconnecting means **shall**:

- Simultaneously disconnect all non-solidly grounded conductors
- Be capable of being locked
- Be one of the following types of disconnects 690.13(E)(1) through (5):
  690.13(E)(1) Manually operated switch or circuit breaker
  690.13(E)(2) Connector as in 690.33(D)(1) or (D)(3) (see page 124)
  690.13(E)(3) Pull-out switch (with required interrupt rating)
  690.13(E)(4) Remote switch that is operated locally and opens when control power is cut
  690.13(E)(5) Device listed or approved for intended operation

Discussion: Recall that a PV system disconnecting means is the borderline between the PV system and other wiring systems. This means that a **dc disconnect in an interactive PV system is not a PV system disconnecting means**. It is merely an equipment disconnect for the inverter. Since the interactive inverter is part of the PV system (only processes PV power), the **PV system disconnect would be an ac disconnect for an interactive system**.

PV system disconnects need to open all non-solidly grounded conductors simultaneously. A PV module connector is an example of a disconnect or isolating device that does not simultaneously disconnect all conductors. It could not be used as a PV system disconnect, but a module connector can be used as an isolation device [690.15].

690.13(E) Informational Note: **"Circuit breakers marked "line" and "load" may not be suitable for backfeed or reverse current**.

This informational note had a big change in the 2020 NEC. Before, in the 2017 NEC and earlier, it was a **shall**, where now it is a **may** in an informational note. There is a big difference between "shall" and "may." The 2017 "shall" means that there is no way you can backfeed something marked "line" and "load," but "may" in this context means that there is a chance in the 2020 and 2023 NEC that you may not be able to backfeed something marked "line" and "load." Thermally operated circuit breakers and fuses cannot tell which way the power is flowing and are generally suitable for backfeed, even if they say line and load on them.

Many people are in jail right now, because they did not know the difference between "shall" and "may," so be sure to read this book carefully and stay safe!

Outline of 690.15 Disconnection Means for Isolating Photovoltaic Equipment

690.15(A) Type of Disconnecting Means
690.15(B) Isolating Device
- 690.15(B)(1) Connector
- 690.15(B)(2) Finger-Safe Fuseholder
- 690.15(B)(3) Isolating Switch Requiring Tool
- 690.15(B)(4) Isolating Device Listed for Application

690.15(C) Equipment Disconnecting Means
- 690.15(C)(1) Have ratings for maximum circuit current, fault current and voltage
- 690.15(C)(2) Simultaneously disconnect non-solidly grounded conductors
- 690.15(C)(3) Externally operable, indicate on or off, capable of being locked under certain conditions
- 690.15(C)(4) Be one of the types in 690.13(E)(1) through (E)(5)

690.15(D) Location and Control
- 690.15(D)(1) Within the equipment
- 690.15(D)(2) In sight from and readily accessible from its equipment
- 690.15(D)(3) Lockable
- 690.15(D)(4) Remote controls that comply with one of the following
  - 690.15(D)(4)(a) Located within the equipment
  - 690.15(D)(4)(b) Lockable and location marked on disconnecting means

690.15 Disconnecting Means for Isolating Photovoltaic Equipment

First, let's discuss the difference between **690.13 PV System Disconnecting Means** and **690.15 Disconnecting Means for Isolating Photovoltaic Equipment**.

690.13 PV System Disconnecting Means, which we just went over, applies to how to separate a PV system from what is not a PV system. You could also consider this the borderline between the NEC definition of a PV system and some other system or special equipment covered outside of Article 690, such as loads in a panelboard, an energy storage system, or a dc coupled battery bank on a 10MW PV system.

690.15 Disconnecting Means for Isolating Photovoltaic Equipment applies to the disconnection of equipment inside the depths of the PV system. Here we can have non-load-break rated equipment, and **non-simultaneous circuit opening, as with module connectors and fuseholders**. We also have equipment disconnecting means, which can be load-break rated and can simultaneously disconnect.

Although Article 690 does not have a definition for "**isolating device**" and perhaps it should, there is an **Article 100 Definition for Isolating Switch**, which can give us some clues. Article 100 is titled Definitions.

> **Switch, Isolating**. A switch intended for isolating an electrical circuit from the source of power. **It has no interrupting rating**, and is **intended to be operated only after the circuit has been opened by some other means**.

The above definition relates to what is often called a **non-load-break disconnect**, or even a connector on the back of a PV module.

To sum it up, PV equipment disconnecting means are always allowed to be **load-break rated** and other times ***not* required to be load-break rated**.

Another term for load-break rated that we may see on equipment is **current interrupting**.

Some inverters or charge controllers have multiple inputs with a single disconnect; this single disconnect with multiple inputs is acceptable.

The purpose of isolating devices is for the safe ability to work on equipment without being exposed to energized conductors, which like Burning Man, may cause serious injury or death. **Isolating devices are not intended to be operated in an emergency condition** since they require that the technician makes sure that the circuit being interrupted does not have current flowing. **It is essentially a maintenance disconnect.**

690.15 states that a disconnect as described in 690.15(A) (see below) shall be provided to disconnect the following equipment from all not-solidly grounded conductors:

- Ac PV modules
- Fuses
- Dc-to-dc converters
- Inverters
- Charge controllers

690.15(A) Type of Disconnecting Means

Disconnects to isolate solar equipment shall be one of the following types:

690.15(A)(1) In Accordance with 690.15(C)

690.15(C) is coming right up on page 92. There are four things here, and they are more the load-break rated types. 690.15(A)(1) is one way of doing things, and there are a few more following.

690.15(A)(2)

Isolating device, which is part of listed equipment where an interlock or similar prevents the opening under load.

690.15(A)(3)

For circuits under **30A** an isolating device in accordance with 690.15(B), which is next.

Turning Off an Inverter with Your Phone

Many devices, such as inverters and PV modules, come with isolating devices or disconnecting means. It is also quite possible that an inverter could be designed with the ability to turn off with the use of a smartphone app. In this case, the smartphone could be used to turn off the inverter and then the connections to the inverter could be unplugged (isolating device) to remove or service the inverter.

690.15(B) Isolating Device

Isolating devices are not required to have an interrupting rating. If the isolating device is not rated for interrupting the circuit current, then it shall be marked with one of the following:

- **Do Not Disconnect Under Load**
- **Not for Current Interrupting** (this phrase is made up of fewer characters and saves label ink)

An isolating device is not required to simultaneously disconnect all current-carrying conductors of a circuit.

For example, if someone is removing or installing a PV module, they will not disconnect or connect positive and negative at exactly the same instant. If they were that quick, they would be making millions playing baseball.

Discussion: PV systems are current-limited, and the equipment of a PV system will have short-circuit currents available that are not much more than operating currents. We have to be sure that the terminals of the equipment side of the disconnect can handle these short-circuit currents.

It is the PV system disconnecting means covered by 690.13 that must have the ability to interrupt higher fault currents coming from batteries and the utility, which are outside of the PV system.

Equipment disconnecting means have to be able to open circuits that have current flowing through them and can be used as an isolating device.

Isolating devices are not required to interrupt current. Isolating devices not rated for current interrupting should not be opened when there is current flowing. If they were opened under load, it could catch fire in rare circumstances, such as in an Article 691 compliant solar farm with no dc arc-fault protection or an old PV system that is pre-2011 NEC without dc arc-fault protection.

690.15(B)(1) through (4)

Following are the only types of isolating devices we can use.

690.15(B)(1) Connector Meeting the Requirements of 690.33

Must be **listed and identified** for use with specific equipment. The primary purpose for this statement was to make it clear that a connector could be used as a disconnect when the application is clearly defined. For instance, consider an MC4 connector that is rated up to 50A and can be installed on circuits up to 1000Vdc. However, on a 1000Vdc series string, if you were to take it apart under even a 5A load, it could start a fire. Take that same connector and put it on the input of a lower dc voltage microinverter and that same connector can easily break the 5A load without damage to the connector or the operator.

Examples are:

- PV module connectors, such as an MC4 or Amphenol connector
- Dc-to-dc converter connector
- Some inverters come with ac cable connectors (more often in other countries)

690.33 Connectors are covered on page 123 of this book.

690.15(B)(2) Finger-Safe Fuseholder

Finger-safe fuseholders are most often used in combiner boxes.

Discussion: When doing voltage or IV curve testing on strings of modules at a dc combiner, we often open the finger-safe fuseholders to **isolate the circuit**. This should only be done after the PV dc combiner disconnect is opened and there is no current going through the circuit. Many solar installers have witnessed an arc show when opening touch-safe fuseholders under load on a system without dc arc-fault protection. Bill always carries a dc clamp ammeter to check for current before opening non-load-break isolating devices.

As we can see in Figure 4.2 on the next page, the finger-safe fuseholder can be opened without touching the fuse. After the fuseholder is opened, both sides of the fuse are de-energized.

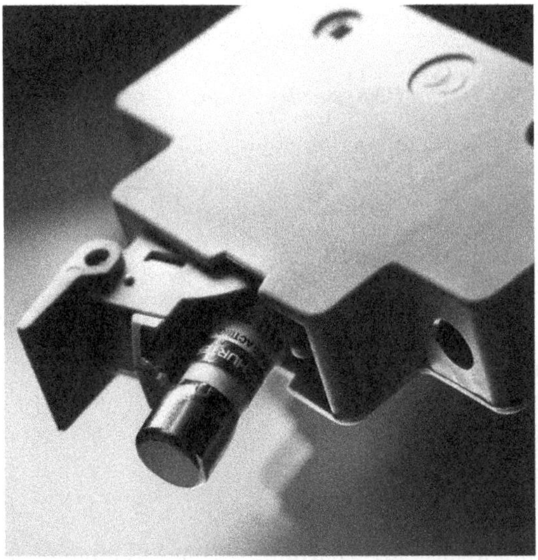

Figure 4.2 Finger-safe fuseholder.
Source: Courtesy Schurter, Inc.

690.15(B)(3) Isolating Device Requiring Tool to Turn Off

For isolating devices that are accessible to non-qualified personnel, these switches require the use of a tool to move the switch to the open position so that it is impossible for someone without the tool to walk by and open the switch under load. Opening one of these switches under load could be extremely dangerous to the operator. Laypeople usually think that turning something off is safe, but when you turn off a non-load-break rated switch under load, it is not safe. This is kind of the opposite of Lockout Tagout (Lockout Tag-on). Here we want to keep the isolating device on instead of off when under load to prevent arcs.

690.15(B)(4) Isolating Device Listed for Application

This option is a catch-all that allows products that have been tested and evaluated for a specific purpose to be used according to how they were evaluated. It may be a combination of several of the options above but configured in a special way that makes it easier to work on the equipment.

690.15(C) Equipment Disconnecting Means

PV equipment disconnecting means shall comply with 690.15(C) (1) to (4)

690.15(C)(1)

Have rating for:

- Maximum circuit current
- Available fault current
- Voltage

690.15(C)(2)

Will simultaneously open current-carrying conductors that are **not solidly grounded**

690.15(C)(3)

- Externally operable
- Will not expose operator to contact with energized parts
- Shall indicate if on or off (you can see that it is on or off)
- If not within sight, or within 10 ft, of disconnecting means or remote operating device shall be capable of being locked in accordance with **110.25 Lockable Disconnecting Means** (capable of being locked with or without lock installed).

690.15(C)(4)

Be one of the disconnect types in 690.13(E)(1) through (E)(5) Type of Disconnect, page 85 (switch, breaker, connector, pull-out switch, remote or listed device)

Discussion: It is an important distinction to note that, in the 2017 NEC and later, we are required to open all **not-solidly**

grounded conductors. Inverters that were previously called grounded inverters in the 2014 NEC that were fuse grounded and were the typical US inverter of the 2000s decade are now called grounded functionally and are *not* solidly grounded. These formerly "grounded" inverters will have their formerly "grounded conductor" opened in the disconnect in the 2017 NEC and later. *Not* disconnecting this former "white wire" that was referenced to ground through a fuse was a requirement of the 2014 NEC and the opposite is true in the 2017 NEC and later. Most large MW scale inverters are still "fuse grounded" (formerly known as grounded) and on new installations both polarities of the disconnect must be opened.

690.15(C) Equipment Disconnecting Means—continued

- If load terminals can be energized in the open (off) position, then the 690.13(B) "line and load energized sign" shall be used **unless 690.33 connectors are used**.
- Connectors between modules and from modules to other equipment are equipment disconnecting means. Fortunately, **we do not need the "line and load energized" label at every solar module**.

2014 NEC and Earlier vs. 2023 NEC (2017 and Later) NEC Inverter/Array Grounding

Comparing NEC requirements for formerly known as "**grounded**" (fuse grounded) inverters:

≤ 2014 NEC

- White grounded conductor
- Opened only single ungrounded conductor per circuit
- Fuses when required only on ungrounded conductor

≥ 2017 NEC

- Positive and negative must not be white unless one is solidly grounded
- Both positive and negative must be opened at disconnect
- Fuses only required on one polarity (positive or negative)

While we are at it, we might as well give the 2023 NEC (2017 and later) and 2014 NEC (and earlier) differences of the formerly known as **ungrounded, transformerless, or non-isolated inverters**, which are also now known as functionally grounded inverters and have the **same rules as all other interactive inverters in the 2023 NEC**.

Formerly known as "**ungrounded**" inverters:

≤ 2014 NEC

- Positive and negative must not be white
- Both positive and negative opened at disconnect
- **Fuses required on positive and negative** when fuses required
- **Must use PV wire**, no USE-2 wire for PV circuits outside of conduit

≥ 2017 NEC

- Positive and negative must not be white
- Both positive and negative opened at disconnect
- **Fuses only required on one polarity** (positive or negative)
- **USE-2 or PV wire both acceptable** for PV circuits outside of conduit

(Note that after the 2020 NEC USE-2 wire has to be dual listed as RHW-2 for this application.)

Now that we know about functionally grounded inverters, perhaps we can say that the 2014 and earlier versions of the Code were instigating **dysfunctional grounding**.

690.15(C) Equipment Disconnecting Means Informational Note

For PV dc disconnects, it is common to put the line side towards the PV. This is more likely to de-energize the load side terminals, blades, and fuses when the disconnect is open.

In addition to the information in this informational note, with a utility interactive system, if you look at the dc and ac disconnects, it is always going to be the inverter side that is safer, so a good idea is to put the line side of the disconnect or other equipment towards the more dangerous side and the load side of the ac or dc disconnect towards the interactive inverter. Think of interactive inverters being smart.

690.15(D) Location and Control

The following are acceptable locations for **isolating devices or equipment disconnecting means** (at least one of the following shall apply):

690.15(D)(1) Within equipment.

690.15(D)(2) Readily accessible from equipment, within **sight and** within **10 feet** of equipment.

690.15(D)(3) Lockable in accordance with **110.25 Lockable Disconnecting Means** (capable of being locked with or without lock installed).

690.15(D)(4) Remote controls which comply with one of the following a. or b.

690.15(D)(4)(a) Disconnect and controls within same equipment.

690.15(D)(4)(b) Disconnect is lockable in accordance with **110.25 Lockable Disconnecting Means** (capable of being locked with or without lock installed). Plus the location of controls are marked on disconnect.

End-of-Chapter Breakdown

Let us break down the different types of disconnecting means as related to PV systems and Article 690 Part III of the NEC.

- Disconnecting means
  - Device(s) that disconnect conductors from supply
  - The three below are all disconnecting means:

(1) PV system disconnecting means
   - Separates PV system from *not* a PV system

(2) Isolating device
   - Non-load-break rated (load-break rated not required)

(3) PV equipment disconnecting means
   - Load-break rated

In other countries that do not abide by the super-safe NEC, it is common for the only load-break rated disconnect in an interactive PV system to be a backfed circuit breaker in a load center. The rest of the system is connected together with connectors so we can say it is "connectorized." These systems are turned off by the backfed breaker and then taken apart while not under load. It appears that the cowboys have gone foreign.

It has been said (although not yet in the NEC) that a well-trained firefighter with baseball experience wielding a fiberglass-handled axe is an exceptional disconnecting means.

5 Article 690 Part IV Wiring Methods and Materials

Article 690 Part IV Wiring Methods and Materials covers the methods and materials specific to wiring PV systems. Much of this material refers to and works with other articles of the NEC, especially the NEC articles within **Chapter 3 Wiring Methods and Materials**, such as **Article 310 Conductors for General Wiring**.

Chapters 5–7 Modify the First Four Chapters of the NEC

This means that if there is a difference between regular wiring methods and Article 690, when we are dealing with a PV system, we will go with Article 690. Same would also go for Articles in Chapter 7, such as 706 Energy Storage Systems, which modifies the first four chapters, including Article 480 Storage Batteries. If Article 690 does not tell us to do something special, then we use the methods found in Chapters 1–4. For example, with ac inverter circuits inside of a house, we can use NMC cable (Romex) and for PV dc circuits on a rooftop, we can use PVC, since the requirements to be in metal are for PV dc circuits inside a building.

Although there are PV-specific wiring methods in 690 Part IV Wiring Methods, all other parts of the 2023 NEC apply to the wiring of PV systems.

DOI: 10.4324/9781003189862-6

690 Part IV Wiring Methods and Materials Sections

690.31 Wiring Methods
690.32 Component Interconnections (short)
690.33 Mating Connectors
690.34 Access to Boxes (short)

Part IV Wiring Methods and Materials Detail

690.31 Wiring Methods
- 690.31(A) Wiring Systems
  - 690.31(A)(1) Serviceability
  - 690.31(A)(2) Where Readily Accessible
  - 690.31(A)(3) Conductor Ampacity
    - Table 690.31(A)(3)(1) Ampacities 105°C to 125°C
    - Table 690.31(A)(3)(2) 105°C to 125°C correction factors
  - 690.31(A)(4) Special Equipment
- 690.31(B) Identification and Grouping
  - 690.31(B)(1) Conductors of Different Systems
    - 690.31(B)(1)(1) Communication circuits allowed together
    - 690.31(B)(1)(2) Inverter output and dc circuits allowed together
    - 690.31(B)(1)(3) Allowing dc circuits together with other circuits
  - 690.31(B)(2) Identification
    - 690.31(B)(2)(a) Polarity identification
    - 690.31(B)(2)(b) Non-solidly grounded identification
  - 690.31(B)(3) Grouping
  - 690.31(B)(3) Exception: If "obvious"
- 690.31(C) Cables
  - 690.31(C)(1) Single Conductor Cable
    - 690.31(C)(1)(a) Dc circuits within array
      - 690.31(C)(1)(a)(1) PV wire or cable
      - 690.31(C)(1)(a)(2) USE-2 and RHW-2
    - 690.31(C)(1)(b) Secured every 24 inches 8 AWG and less
    - 690.31(C)(1)(c) Secured every 54 inches over 8 AWG (over 6 AWG)
  - 690.31(C)(2) Cable Tray
    - 690.31(C)(2)(1) Single layer for single conductor cables
    - 690.31(C)(2)(2) Circuits bound together in other than single layer

690.31(C)(2)(3) Sum of diameters of single conductor cable not exceed width

690.31(C)(3) Multiconductor Jacketed Cables

690.31(C)(3)(1) In raceways on or in buildings (not roofs)

690.31(C)(3)(2) Not in raceways

690.31(C)(3)(2)(a) Sunlight resistant

690.31(C)(3)(2)(b) Protected or guarded if subject to damage

690.31(C)(3)(2)(c) Follow surface of support structures

690.31(C)(3)(2)(d) Secured intervals 6 feet or less

690.31(C)(3)(2)(e) Secured within 24" of connectors or enclosures

690.31(C)(3)(2)(f) Marked direct burial when buried in earth

690.31(C)(4) Flexible Cords and Cables Connected to Tracking PV Arrays

Table 690.31(C)(4) Minimum PV Wire Strands (for Flexible Cords and Cables)

690.31(C)(5) Flexible, Fine-Stranded Cables

690.31(C)(6) Small-Conductor Cables

690.31(D) Direct-Current Circuits on or in Buildings

690.31(D)(1) Metal Raceways and Enclosures

690.31(D)(2) Marking and Labeling

690.31(D)(2)(1) Exposed raceways, cable trays, etc.

690.31(D)(2)(2) Pull and junction boxes

690.31(D)(2)(3) Conduit bodies

690.31(E) Bipolar Photovoltaic Systems

690.31(F) Wiring Methods and Mounting Systems

690.31(G) Over 1000 Volts Dc

690.31(G)(1) Not on 1- and 2-family dwellings

690.31(G)(2) Not within habitable buildings

690.31(G)(3) Exterior of building requirements <10' above grade & < 33'

690.32 Component Interconnections

690.33 Mating Connectors

690.33(A) Configuration

690.33(B) Guarding

690.33(C) Type

690.33(D) Interruption of Circuit

690.33(D)(1) Rated for Interrupting

690.33(D)(2) Requires tool to open and marked
690.33(D)(3) Connectors can be MLPE disconnects if listed
690.34 Access to Boxes

690.31 Wiring Methods

It is amusing when we see 690 Part IV **Wiring Methods and Materials** / 690.31 **Wiring Methods** / 690.31(A) **Wiring Systems**. We see a lot of repetition that emphasizes the **importance** of our **wiring method and materials systems**!

690.31(A) Wiring Systems

690.31(A)(1) Serviceability

Sufficient length of cable shall be provided in enclosures for replacement.

Discussion: Since your inverters and equipment will likely be replaced at some point, we need to make it easy for that future person, who may be your future self (with a different set of atoms).

> **300.14 Length of free conductors at outlets, junctions and switch points** says that we need **6 inches of free conductor**. It also says that, for boxes measuring fewer than 8 inches in any dimension, we only need the wire to be able to extend 3 inches outside an opening. Additionally, if the wires are not spliced in the box, then this requirement does not apply.

690.31(A)(2) Where Readily Accessible

PV dc circuits over 30V in **readily accessible locations** shall be one of the following:

(1) Guarded
(2) In MC cable (Article 330 Metal-Clad Cable: Type MC)
(3) In raceway (does not require metal raceway here; however, requirement for being in metal is for dc circuits inside of buildings. See 690.31(D)(1) on page 119).

690.31(A)(3) Conductor Ampacity

Table 690.31(A)(3)(1) shall be used for correcting ambient temperatures over 30°C for conductors with 105°C and 125°C rated insulation. The temperature rating of the insulation is theoretically how hot the wire can get after running the current in the ampacity table through it under given conditions. At this point, nobody is using 105°C and 125°C wire; however, Bill put this in the NEC, since he is an NEC futurist and copper cost reductionist. For proper wire sizing, see Chapter 12 Wire Sizing. Table 690.31(A)(3)(1) is essentially an extension of Table 310.16, but with higher temperature rated conductors.

Table 690.31(A)(3)(1) in the 2023 NEC was Table 690.31(A)(a) in the 2020 NEC and 690.31(A) in the 2017 NEC. In the 2017 NEC, the temperature rating of the wires included temperatures covered in the 310 Tables. In the 2020 and 2023 NEC, this table covers 105°C and 125°C insulation temperatures only, which are very hot! Perhaps we are preparing for a Venus-like climate (this is not a political statement).

> Table 310.15(B)(2)(a) Ambient temperature correction factors based on 30°C in the 2017 NEC was renamed Table 310.15(B)(1) in the 2020 NEC and then 310.15(B)(1)(1) in the 2023 NEC. **The 310 Tables** have been seeing a lot of name changes lately, but the content remains the same. See page 275 for details. Perhaps that Code Making Panel wants more attention.

690.31(A)(3) Continued

Table 690.31(A)(3)(2) shall be used for determining ampacities of conductors rated for 105°C and 125°C. This table can be viewed as an extension of the table now known as **Table 310.16**. [From the 2011 to the 2017 NEC this table was temporarily named 310.15(B)(16).]

Although conductors with a higher temperature rating can carry more current, they will still have the same voltage drop as conductors with a lower temperature rating. Since PV prices are getting so low, at this rate, they will pay us to take it in a few more

Table 5.1 690.31(A)(3)(1) correction factors for 105°C and 125°C rated wire (ambient temperature correction factors for temperatures over 30°C)

Ambient temperature (°C)	*Temperature rating of conductor*	
	105°C (221°F)	*125°C (257°F)*
30	1.00	1.00
31–35	0.91	0.94
36–40	0.82	0.88
41–45	0.71	0.82

* This table goes up to 120°C (248°F) ambient temperatures in the NEC
Source: Courtesy NFPA.

Code cycles, so why not save money on smaller wire? If you do not see Bill's name on the next edition of this book, it means the copper industry has taken him out for reducing the amount of copper required in a PV system. Sean hears the OCPD manufacturers also have put out a hit on him.

690.31(A)(4) Special Equipment

Wiring methods listed for PV systems may be used, in addition to wiring methods in the Code.

Permitted Wiring Methods

- Raceway wiring methods in the NEC
- Cable wiring methods in the NEC
- Other wiring systems specifically **listed for PV arrays**
- Wiring that is part of a **listed system**

The following are comments on different wiring systems:

- Raceway wiring methods in the NEC
  - Raceway wiring methods are found primarily throughout Chapter 3 of the NEC. Chapter 3 is titled **Wiring Methods and Materials**.

Table 5.2 690.31(A)(3)(2) Ampacities of 105°C and 125°C rated wire not in free air

	Ampacities	
AWG	*105°C(221°F)*	*125°C(257°F)*
14	29A	31A
12	36A	39A
10	46A	50A
8	64A	69A
6	81A	87A
4	109A	118A
3	129A	139A
2	143A	154A
1	168A	181A
1/0	193A	208A
2/0	229A	247A

Source: Courtesy NFPA.

  - EMT is often the raceway method of choice and is covered in **Article 358 Electrical Metallic Tubing: Type EMT**.
- Cable wiring methods in the NEC
  - Cable wiring methods are also found in Chapter 3 of the NEC.
  - USE-2, a popular cable, is covered in Article 338, Service-Entrance Cable: Types SE and USE. The -2 of USE denotes that it is 90°C rated.
  - PV wire is another wiring method that was specifically required in the 2014 NEC, but is an option along with USE-2/RHW-2 in the 2023 NEC.
- Other wiring systems specifically **listed for PV arrays**
  - A cable manufacturer could design a cable that was superior to PV wire, get a certifying agency to list it, and now we have another cable listed for PV arrays.
- Wiring that is part of a **listed system**
  - While this listed system could be a wiring system for rapid shutdown, it could be any special cable used as part of a listed system. The Enphase microinverter cable fits under this "listed system" category.

690.31(A)(4) INFORMATIONAL NOTE

See **110.14(C) Temperature Limitations** under **110.14 Electrical Connections** or see how to do this in Chapter 12 Wire Sizing of this book, which is easier to understand than the NEC (why we wrote it). Also explained on page 44.

690.31(B) Identification and Grouping

There was a little bit of rearranging here in the 2023 NEC, including turning things into lists.

690.31(B)(1) Conductors of Different Systems

Unless allowed in the important exceptions below or allowed in equipment listing, then PV circuits shall not occupy the same equipment wiring enclosure, cable, or raceway as other non-PV systems or inverter output circuits, unless separated by barrier or partition.

Partitions are available with some enclosure products such as gutters. **Most jurisdictions will require a partition to be an accessory included in the listing of the enclosure.**

690.31(B)(1) EXCEPTION (SOMETIMES THE EXCEPTION IS MORE COMMON THAN THE RULE)

If all conductors and cables have an insulation **voltage rating that is at least that of the highest** maximum circuit voltage [690.7(A) defines maximum circuit voltage], then the **following** 1, 2, & 3 **shall be permitted**:

690.31(B)(1)(1) Power limited circuits **Multiconductor jacketed cables** for remote **control**, **signaling**, or **power limited circuits** are permitted in the same enclosure/cable/raceway as PV dc circuits (assuming all circuits serve the PV system).

Discussion: Think of this as your data, rapid shutdown signaling, and CT cables. You just need to make sure the insulation is rated for the higher voltage. 1500V CAT5 network cables are available by the way.

Example: If there is a 600V circuit and a 24V circuit in the same jacketed cable/raceway/enclosure, then the insulation around the 24V circuit has to be rated for at least 600V.

690.31(B)(1)(2) Inverter output circuits **Inverter output circuits (ac) can occupy** the same box or wireway as **PV dc circuits**, that are identified and grouped [which is required anyway according to **690.31(B)(2) Identification** and **690.31(B)(3) Grouping** and coming right up].

Discussion: Since identification and grouping is already required, then we almost do not need to mention grouping and identification.

690.31(B)(1)(3) Inverter output circuits and non-PV systems Dc circuits using multiconductor jacketed cable, MC cable, or listed wiring harnesses that are identified for the application and permitted to occupy the same wiring method as inverter output circuits and non-PV systems.

Discussion: Here we can even combine non-PV circuits in the same wiring method, so as we said, the exception is more of a rule.

690.31(B) also tells us we need to comply with both 690.31(B)(1) Identification and Grouping and 690.31(B)(2) Grouping.

690.31(B)(2) Identification

PV system dc circuit conductors shall be **identified** at accessible points of

- Termination
- Connection
- Splices
- Exception: If evident by spacing or arrangement, further identification not required

Means of identifying PV system circuit conductors:

- Color coding
- Marking tape
- Tagging (probably not spray-painted gang tagging)
- Other approved means in 690.31(B)(2)(a) and (b) coming right up.

690.31(B)(2)(A) OTHER APPROVED MEANS OF IDENTIFICATION

Using color for polarity identification shall be with a permanent method, **such as**

- Labeling
- Sleeving
- Shrink tubing

690.31(B)(2)(B) AND ONE OTHER APPROVED MEANS OF IDENTIFICATION

Non-solidly grounded positive conductors shall include either:

- Imprinted plus sign "**+**"
- The word **POSITIVE or POS** durably marked and not green, white, or gray

Non-solidly grounded negative conductors shall include either:

- Imprinted negative sign "–"
- The word **NEGATIVE or NEG** durably marked and not red, green, white, or gray

Note: only solidly grounded conductors shall be marked in accordance with **200.6 Means of Identifying Grounded Conductors**, which is where you find the white wire for the neutral rule and alternatives, such as gray.

Discussion: **Solidly grounded systems** in the 2023 NEC (2017 NEC and after) are rare systems that are ***not* fuse grounded**. An example of a solidly grounded PV system according to the 2023 NEC is a **direct PV well pump**, where the negative conductor is solidly connected to a grounding system that includes a grounding electrode (often the well casing). Grounding through a fuse is not solidly grounding.

These rare and solidly grounded conductors will most likely be marked white according to 200.6 (the marking could be grey, three white stripes, or three grey stripes). **Fuse-grounded current-carrying conductors** that operate at zero volts to ground **are not considered to be meeting the requirements of 200.6** and **should *not* be identified as white, as they were in the 2014 NEC and earlier versions**. The ac side of an inverter is most often solidly grounded, just like anything else ac.

The **690.31(B)(2) Identification Exception** tells us that if the identification of conductors is evident by spacing or arrangement, then we do not need other forms of identification.

690.31(B)(3) Grouping

If the ac and dc conductors of PV systems occupy the same

- **Junction box** or
- **Pull box** or
- **Wireway**

then the ac and dc conductors of each system shall be grouped separately by **cable ties** or similar every **6 ft**.

The **690.31(B)(3) Grouping Exception** tells us that the grouping **does *not* apply** if the circuit **enters from a cable or raceway that is unique and obvious**.

(What is **obvious** to a qualified person is oblivious to someone who has not read this book.)

690.31(C) Cables

PV wire or cable and distributed generation (DG) cable shall be listed. The PV Wire listing is UL 4703 and DG cable is UL 3003.

UL listing numbers in this industry are related.

The UL **listing for PV is UL 1703 (or 61730, see below)**. If you change the 03 to 41, then you have the **inverter, charge controller, and dc-to-dc converter UL listing 1741**. If you change the first 1 to a 3 in UL 1741, you get UL 3741, which is the rapid shutdown listing we call PV Hazard Control Systems (**UL 3741 PVHCS**). If you change the first digit in UL 1703 from a 1 to a 2, then you get the **racking listing of UL 2703**. If you change the first digit in UL 1703 from a 1 to a 4, then you get the listing for **PV wire, which is UL 4703**. If you change the first digit in UL 1703 from a 1 to a 6, then you get the listing for **PV connectors, which is UL 6703**. For DG cable, you change the 17 in UL 1703 to a 30, then you get the listing for **DG cable, which is UL 3003**.

As of December 2020, there became a new PV UL listing in town, which is **UL 61730**, rather than **UL 1703**. This is because the listing was harmonized with the Euro/

international listing, so the PV manufacturers no longer have to do two separate tests for a globally used PV module. All they did was add a 6 to the beginning of 1703, which is apparently some **International Electrotechnical Commission (IEC)** thing; however, a 6 before **UL 1703** was already taken by the IEC 61703 Mathematical Expression for Reliability (which is unrelated to PV), so the 0 and the 3 were transposed by some high paid transposing expert (numerologist), and now we have the well thought out **UL 61730 listing**. If something is listed as **UL 1703** you can still use it, but products first made after 2020 will do the **UL 61730** test.

690.31(C)(1) Single-Conductor Cable

Single conductor cable in **exposed outdoor** locations in **PV dc circuits within PV array** shall comply with 690.31(C)(1)(a), (b) and (c):

690.31(C)(1)(a) There are two wire types you can choose from:

690.31(C)(1)(a)(1) **PV wire or cable** (has non-standard diameter, so if you are doing conduit fill calculations, you have to get the wire dimensions from the manufacturer).

690.31(C)(1)(a)(2) **Dual listed as USE-2/RHW-2** (almost always USE-2 is dual listed as RHW-2)

Discussion: Dual listing USE-2/RHW-2 was new in the 2020 NEC. Most USE-2 wire already is dual listed as RHW-2, so it would be more difficult to find USE-2 wire that was not also RHW-2 wire. USE-2 wire on its own cannot go inside of conduit inside a building, but when it is dual listed it can go inside of buildings on the coattails of RHW-2. The RHW designation includes the building wire fire ratings required for conductors in buildings. PV wire also has this same fire designation and is usually also marked RHW-2.

690.31(C)(1)(b)

- 8 AWG and smaller shall be supported every 24".
- PV wire is permitted wherever RHW-2 is permitted (like in conduit).
- Exception for Article 691 Large-Scale PV Electric Supply Stations (see page 171) where you can do various things that

conflict with 690 with an engineered design. 691.4 defines 691 compliant, including requirement to be over 5MW ac. For these large PV systems, you can secure at intervals determined by the engineer(s).

Discussion: In the 2017 NEC and earlier there was no 24" requirement and we looked to 338.10(B)(4)(b) and 334.30, which let us support the conductors every 4.5 ft. Some installers would like to have the distance be at least the width of the module, which is usually about 40 inches.

690.31(C)(1)(c)

- Larger than 8 AWG shall be supported every 54".
- This is **new** in the 2023 NEC.

USE-2 and PV Wire

USE-2 cable and PV wire are the two most common wiring methods for connecting PV modules to each other and for connecting PV modules to anything else. These wiring methods are commonly installed under PV modules and do get exposed to sunlight. It is interesting that USE stands for Underground Service Entrance, and **has been tested for exposure to sunlight**. (Not usually much sunlight underground unless you consider neutrinos sunlight.) The **-2 suffix** is one way of indicating that something is **90°C rated**.

In previous versions of the NEC, PV wire was specifically required for what was formerly known as “ungrounded” PV arrays in the 2014 NEC, but **now there is no specific requirement to use PV wire**. PV wire may be better than USE-2 wire and has been tested with more UV light, but USE-2 cable may be less expensive.

When the PV industry was originally trying to get the code-making panel to accept “ungrounded” inverters, they used similar wiring methods to systems with double insulation in Europe. PV wire is often the wire of choice of PV module manufacturers since it is acceptable everywhere in the world and with every version of the NEC.

USE-2 cable, and PV wire, are often colored black and **work better when colored black**, since black is the color of carbon black, which is a pigment that helps with UV resistance by preventing UV light from penetrating. This is also why black cable ties are more UV resistant (and why light skin can lead to deadly melanomas). When sourcing black cable ties, use **Nylon 6 cable ties** when contacting everything except galvanized steel. **Nylon 12 is necessary for any cable ties contacting galvanized steel**. One last item for cable ties: the NEC and UL standards **require that the cable ties be rated "Type 21C"** where used for cable support. This Type 21C designation will typically be printed on the package enclosing the ties.

There was a time when some inspectors did not read **200.6 Means of Identifying Grounded Conductors** and they believed that white USE-2 cable was required. (There are other ways of identification, such as marking.)

Some installers have used red USE-2 cable to indicate a positive conductor and as the red faded, people would see white, and think that the formerly red wire was a white grounded conductor.

Most, if not all PV wire is also marked RHW-2 since it has to pass all the same tests as RHW-2. This rating of RHW-2 also allows the conductor to be installed inside buildings. USE-2 without the RHW-2 marking is not permitted, because it may not have been tested for the necessary fire and sunlight ratings.

690.31(C)(2) Cable Tray

There were some changes here, and some jumping around the NEC is required, such as jumping from Articles 690 to **392 Cable Trays** and **310 Conductors for General Wiring**. In previous versions of 690, what was missing was what to do about conductors smaller than 1/0 in cable trays and then some wise guy, Dave Click PE, pointed it out and caused us to make this book longer which comes down to killing trees in the name of saving the planet. Dave is a proofreader of this book, so if there are any mistakes, they are all his fault.

Not changed in the 2023 NEC were the following requirements:

- Support cables every 12".
- Secure the cables every 54".
- PV wire, PV cable or DG cable without cable tray rating can go in cable tray outdoors.

Now for the new rules which **apply to single-conductor PV wire smaller than 1/0**:

- In uncovered cable trays the adjustment factors for 1/0 in 392.80(A)(2) are used:
    - Article 392 Cable Trays
    - 392.80 Ampacity of conductors
    - 392.80(A) Ampacity of Cables, Rated 2000V or Less, in Cable Trays
    - **392.80(A)(2)** Single-Conductor Cables
- **392.80(A)(2)** starts out not being specific to 1/0 and larger. It does tell us some important things that we should do when using cable trays, such as:
    - Obey **310.14(A)(2) Selection of Ampacity**, which is where we select the lowest ampacity in a circuit, unless that lower ampacity is **10' or 10%** of the circuit length, whatever is less. This is a common wire sizing rule, that you should be familiar with. See page 275.
    - Do not obey **310.15(C)(1) More than 3 Current Carrying Conductors**, which are the familiar adjustment factors for greater than 3 current carrying conductors, which any wire sizing geek knows inside out. (Page 276.)
    - Then **392.80(A)(2)** tells us that **for our single conductor cables**, or when our cables are cabled together (such as quadruplexed), we then need to follow a through d.
    - **392.80(A)(2)(a)** is for huge 600 kcmil and larger cables and we are told to use 70% of ampacities in Tables 310.17 and 310.19.
    - **392.80(A)(2)(b)** is for 1/0 through 500kcmil and this is what 690.31(C)(2) is referring to and what we are mostly looking for, since we are supposed to use what *392.80(A)(2)(b)* says

for 1/0 cables, for cables smaller than 1/0. (Note that (c) and (d) following also apply to 1/0). Now what it says here are two different things.

- **Use 65%** of ampacities in Tables 310.17 and 310.19.
- **Use 60%** if the cable trays are continuously covered and covers unventilated for 6' or more.

Tables 310.17 and 310.19

What we are mostly familiar with is **Table 310.17 Ampacities of Single-Insulated Conductors in Free Air** which is based on 30°C ambient temperature, and then using table 310.15(B)(1)(1) Ambient Temperature Correction Factors Based on 30°C to correct for hotter temperatures. The NEC does have tables based on 40°C, such as **310.19 Ampacities of Single-Insulated Conductors in Free Air which is based on 40°C.** I have always been taught to ignore ampacity tables based on 40°C. I recall someone on the Mike Holt team crossing out the page for a 40°C ampacity table. We like to treat our NEC books as sacred and do not put marks on them.

So just use 310.17 for cable trays and use a special cable tray derating factor, which is probably going to be 65% or 60%.

- **392.80(A)(2)(c)** is for 1/0 and larger, so we use it for smaller than 1/0 for PV. When single conductors are installed in cable tray with spaces between the cables that are at least one cable diameter. There are no derating factors here, so you get 100%!
- **392.80(A)(2)(d)** is for 1/0 and larger, so we use it for smaller than 1/0 for PV in these cases. If single conductors are installed in triangular or square configuration, with a free air space between them of at least 2.15 times the conductor diameter of the largest conductor, then there is no derating factor, so 100%, just like we scored on our NABCEP exams!

Now back from our detour to Articles 392 and 310 and to 690.31(C)(2) where it says:

Where single-conductor PV wire smaller than 1/0 AWG is installed in ladder ventilated trough cable trays, the following (three) shall apply.

690.31(C)(2)(1) Single conductors installed in 1 layer
690.31(C)(2)(2) Conductors bound in circuit pairs can be installed in other than a single layer
690.31(C)(2)(3) Sum of diameters of conductors shall not exceed cable tray width

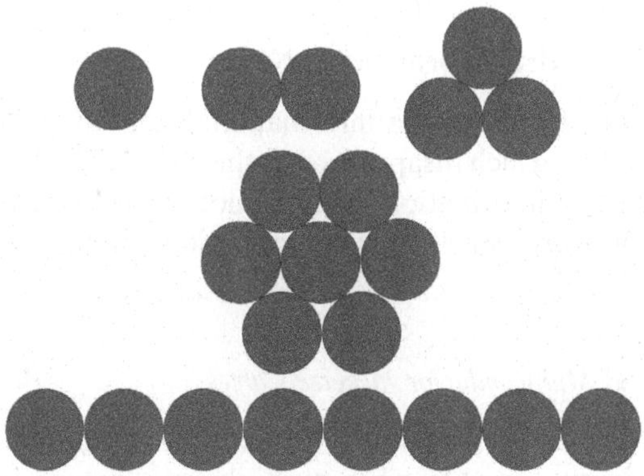

Figure 5.1 Conductors in different configurations.
Source: Sean White.

Figure 5.1 also relates to Chapter 9 Table 1 Percent of Cross Section of Conduit and Tubing for Conductors and Cables. It is all about geometry and where you can fit the circles.

Summing up 690.31(C)(2) Cable Trays:

1. We support our cables every 12" and secure them every 54" (or less).
2. Instead of using Table 310.15(C)(1) Adjustment Factors for More than 3 Current Carrying Conductors when using cable trays, we find other rules and adjustment factors, such as a 65%, 60% or 100% (no) factor in Article 392 Cable Trays, depending on the cooling effects of spacing and air flow.

PV Wire or Cable and DG Cable Diameter

If you are looking for the diameter of your PV wire, take note that:

> PV and DG Wire/Cable has a non-standard diameter (it can vary from one manufacturer to the next), so we cannot use Annex C and we have to get the wire diameter from the manufacturer and use that along with Chapter 9 Tables 1, 4 and 5 for determining how many wires fit in conduit.

PV and the NEC Informational Note:

This quote above was an informational note in 2020 NEC 690.31(C)(2), which disappeared in the 2023 NEC, but still is good enough information for this book, which is really just one big and exciting 300 page informational note.

690.31(C)(3) Multiconductor Jacketed Cables

Although not defined in the NEC, a multiconductor jacketed cable is simply a cable with multiple conductors in it that also has a jacket around the cable. A common example of a multiconductor jacketed cable would be that yellow or white NMC cable (a.k.a. Romex) in the studs of your walls. You can also have DG cable and service entrance cables that fit this definition, or even your ethernet cable going to a router or energy storage system.

If said cables are part of a listed assembly, then install them according to the instructions. You always follow instructions for listed equipment because that is how it was tested to be safe. Also, if something bad happens in the system and instructions are not followed, the lawyers will make a lot of money.

A microinverter cable is an example of this kind of cable that is part of a listed system.

If the cables are **not part of a listed assembly**, then you can still install these **multiconductor jacketed cables** in accordance with the cable's listing and in accordance with the following (1) through (2) a to f.

Table 5.3 NEC Table 690.31(C)(4) Minimum PV Wire Strands

PV wire AWG	*Minimum strands*
18	17
16–10	19
8–4	49
2	130

Source: Courtesy NFPA.

- 690.31(C)(3)(1) If **on or in buildings** then **in raceway**
- 690.31(C)(3)(2) If **not in raceways**, then a to f
  - 690.31(C)(3)(2)(a) Marked **sunlight resistant if exposed outdoors**
  - 690.31(C)(3)(2)(b) **Protected or guarded when exposed to damage**
  - 690.31(C)(3)(2)(c) **Closely follow the surface or support structure**
  - 690.31(C)(3)(2)(d) **Secured** no less than every **6 ft**
  - 690.31(C)(3)(2)(e) Secured no more than **24 inches from connectors** or entering **enclosures**
  - 690.31(C)(3)(2)(f) Marked **direct burial if buried**

To sum up multiconductor jacketed cables: If it is **inside, it needs to be in a raceway** and if it is **outside,** then it needs to be **in a raceway or meet all six things** above a through f, **or** according to instructions if **part of a listed PV assembly**.

690.31(C)(4) Flexible Cords and Cables Connected to Tracking PV Arrays

The following shall apply to flexible cords and cables connected to tracking PV arrays:

- Identified as **hard service cord** or **portable power cable**
- Suitable for **extra-hard usage**
- Listed for outdoor use
- Water resistant
- Sunlight resistant
- Stranded copper permitted to be connected to moving parts of tracking PV arrays in accordance with **Table 690.31(C)(4)** (see below)

- Comply with **Article 400 Flexible Cords and Cables**
- Allowable ampacities in accordance with **400.5 Ampacities for Flexible Cords and Cables**

Article 400 Flexible Cords and Flexible Cables Discussion: Article 400 contains different tables to be used for flexible cords and cables, which are used for stationary conductors, including various ampacity tables, much **like the Article 310 tables**, but more "flexible" (pun) in **400.5 Ampacities for Flexible Cords and Flexible Cables**.

Table **690.31(C)(4) Minimum PV Wire Strands** Discussion: Tracking arrays have moving parts that may bend wires back and forth 365 times per year, which can be 10,000 times or more in the life of a 30-year-old PV system. Wire with not enough strands will strain-harden and break.

Table **690.31(C)(4) Minimum PV Wire Strands** does not have in its title that it is specific for tracking arrays and, if taken out of context, someone might think that it would apply with all PV wire or all PV system wiring. **This table is only required for PV wire connected to moving parts of PV tracking arrays.** For other types of wire connected to tracking systems, you would look in **Article 400 Flexible Cords and Cables, where PV wire is not covered.**

The **ampacities in 400.5 and Tables 400.5(A)(1) and 400.5(A)(2) differ from the ampacities typically used for conductors in free air** in Table 310.17. These cables also have different ampacities depending on how many conductors are contained within the cable.

The ambient temperature correction factors in Table 310.15(B)(1)(1) [formerly 310.15(B)(1) in 2020 NEC or 310.15(B)(2)(a) in 2017 NEC] do apply.

It is interesting to note that Table 400.5(A)(3) Adjustment Factors for More than Three Current Carrying Conductors in a Flexible Cord or Flexible Cable is exactly the same as Table 310.15(C)(1) Adjustment Factors for More Than Three Current Carrying Conductors. What a waste of ink! (Now we're doing it too.)

690.31(C)(5) Flexible, Fine-Stranded Cables

Flexible, fine-stranded cables shall only be terminated with lugs, devices, or connectors in accordance with **110.14 Electrical Connections**.

110.14 Electrical Connections includes information on (A) Terminals, (B) Splices and (C) Temperature Limitations.

In 110.14(c) it says that **conductors with temperature ratings higher than specified for terminations may be used for** ampacity adjustment, correction, or both. This is something that is more of the complicated part of wire sizing. **When you are correcting for ampacity with adjustment and correction factors, you do not take terminal temperatures or the 125% continuous current correction factor into consideration.** You do take the terminal temperature limits and the 125% required ampacity for continuous current into consideration without the adjustment and correction factors. You can read more about wire sizing in Chapter 12 of this book beginning on page 287 to see how it is done in practice, and it is also covered somewhat when we discuss 690.8(B) beginning on page 41. Just because something is in the NEC and you have to do it, it does not always mean that it has to make sense. With some things, once you understand how it does not make sense, then you understand it.

Flexible, Fine-Stranded Cables, Formerly Known as 690.31(H)

Although 690.31(H) was moved to the cable section as 690.31(C)(5), the NEC already has provisions for using proper terminations for the application in **110.14 Electrical Connections**, but we are reminded about it here since many installers tend to forget this part of the Code.

Flexible, fine-stranded cables shall be terminated only with terminals, lugs, devices, or connectors in accordance with 110.14.

Flexible, fine-stranded cables are easy to bend, but are more difficult to terminate. Inexperienced installers in the past have used typical screw terminals that are not meant for flexible, fine-stranded cables. In these cases, the connection will often become loose, resistance will rise and heat will be generated, which can be a fire hazard.

PV installers in the past used batteries and liked to use flexible, fine-stranded cables, since they would not overstress battery terminals. Most battery cables now use fine-stranded cables with the proper terminations. Batteries are no longer part of a PV system as of the 2017 NEC.

PV arrays that track the sun do require flexible cables and installers of these cables should be aware of the requirements for using proper terminal procedures and equipment.

690.31(C)(6) Small Conductor Cables

16 American Wire Gauge (AWG) and 18 AWG single-conductor cables are **permitted for module interconnections** if they:

- Meet ampacity requirements of **400.5 Ampacities for Flexible Cords and Cables.**
- Comply with correction and adjustment factors from **Section 310.14 Ampacities for Conductors Rated 0–2000 Volts.**
  - **310.14 does point to the famous tables used for wire sizing in 310.15 among other things, such as 310.15(B)(1)(1) Ambient Temperature Correction Factors Based on 30°C and 310.15(C)(1) Adjustment Factors for More than Three Current Carrying Conductors.**

Note that 310.14 does not direct you to Tables 310.16 and 310.17 and instead we use the ampacity tables in Article 400 for Flexible Cords and Cables, even though these small conductors are not necessarily rated as flexible cords or cables.

Discussion: It may have been **unthinkable to use 16 AWG or 18 AWG wire for PV systems** when PV was more expensive; however, with the advent of falling prices, it is now thinkable and more likely in large projects using thin film modules with lower than typical crystalline silicon PV module current ratings. 310.10(A) Minimum Size of Conductors says that we cannot use any copper conductor smaller than 14 AWG unless stated elsewhere in the Code. 690.31(C)(6) is officially elsewhere in the Code.

Did you know that a MW at a MV is 1A?

690.31(D) Direct Current Circuits on or in Buildings

Wiring on or in buildings must comply with 690.31(D)(1) and 690.31(D)(2) below.

690.31(D)(1) Metal Raceways and Enclosures (PV dc circuits inside of buildings)

PV dc circuits inside a building **that exceed 30V or 8A** shall be inside of:

- Metal raceway (like EMT Article 358)
- MC cable (Article 330, Metal-Clad Cable: Type MC). We are also directed to 250.118(10)(b) or (c), which tells us how MC cable can be also used as an equipment grounding conductor:
  - **250.118(10)(b)** Combined metallic sheath **and** uninsulated grounding conductor of interlocked metal tape of MC cable that is listed and identified as equipment grounding conductor (EGC).
  - **250.118(10)(c)** metallic sheath **or** combined metallic sheath **and** EGC of smooth corrugated tube-type MC cable that is listed and identified as EGC.

690.31(D)(1) Exception to Metal Raceways and Enclosures *(PV dc circuits inside of buildings)*

PV Hazard Control Systems (PVHCS) in accordance with 690.12(B)(2)(1), (UL 3741 listed PVHCS, see page 69) shall be permitted to use non-metallic or use other than MC cables at the point of penetration of the building.

Things Removed in the 2020 NEC to Take Note Of

This "**inside of metal**" requirement **used to be** in the 2017 NEC and earlier, and **only applied up to the first readily accessible dc disconnect**. This is no longer the case. Dc parts of a PV system in a building require metal even after the first dc disconnect, unless part of listed UL 3741 PVHCS.

Also, take note that the special requirements that were in the 2017 NEC regarding **PV circuits embedded in building surfaces** were removed in the 2020 NEC. Perhaps due to 690.12, the roof is safer, and these requirements are no longer needed.

Removed in the 2023 NEC were requirements for **PV dc circuits in buildings** that were using flexible wiring methods, such as FMC and MC cable, that required protection, such as **guard strips**.

690.31(D)(2) Marking and Labeling

PHOTOVOLTAIC POWER SOURCE or **SOLAR PV DC CIRCUIT**

Unless located and arranged so the purpose is evident, this wording above **shall be marked** on the following wiring methods and enclosures that contain PV system **dc** circuits **on or in buildings (not ac! Only on or in buildings!)**:

690.31(D)(2)(1) Exposed raceways, cable trays and other wiring methods
690.31(D)(2)(2) Covers or enclosures of pull and junction boxes
690.31(D)(2)(3) Conduit bodies with unused openings

Label shall be:

- Every 10 ft
- Every section separated by:
  - Enclosures
  - Walls
  - Partitions
  - Ceilings
  - Floors

The **label specifications** shall be:

- CAPITALIZED
- 3/8-inch height minimum
- White letters
- Red background
- Label shall be suitable for the environment used

Note: SOLAR PV DC CIRCUIT is made up of fewer letters, and takes up less space and resources, so it is better for the environment and helps save the planet.

690.31(E) Bipolar Photovoltaic Systems

Discussion: Bipolar PV systems have a benefit that voltage, for the purposes of the Code, can be considered as being voltage to ground rather than the maximum voltage between any two conductors.

Analogy:

Your house in the US is wired at 120/240V split phase. This means that you will not have more than 120V to ground in your house; however, you can have an electric dryer that gets all the benefits of 240V. This is because your house is bipolar in a way and is grounded and has a grounded conductor right in the middle of that 240V.

A bipolar system has two **monopole circuits** (typically groups of strings), one positively grounded and the other negatively grounded. This type of system can potentially give the array 25% of the power losses (75% savings) due to voltage drop at double the voltage and half the current that it otherwise would have.

The wiring rules with these systems state that we cannot have conductors next to each other that could have voltages greater than Code allows for the wiring method and the conductors. If these conductors are in the same location, they need to be able to handle these bipolar hot-to-hot voltages than are double of what the voltage to ground can be.

In other words, going bipolar does not give a designer the right to have two wires next to each other that have voltages greater than that for which they are rated.

At the time of the writing of the 2023 NEC, we did not see much in the way of bipolar systems. It is invigorating to think that we could have a "1500V to ground" ground-mounted system that would have conductors that could measure 3000V to each other and still be NEC compliant! That could be efficient with the low relative currents. It is rumored that some bipolar plans may be in the works.

Bipolar systems were much more popular just over a decade ago when the maximum system voltage for all PV systems was 600V rather than 1500V.

690.31(F) Wiring Methods and Mounting Systems

Roof mounting of PV systems does not need to comply with 110.13.

110.13 Mounting and Cooling includes 110.13(A) Mounting and 110.13(A) says that equipment shall be "firmly secured in place." Since many PV systems are ballasted and held in place by weight and aerodynamics, **690.31(F) is formally allowing ballasted systems.**

We are required by 690.31(F) to use an approved method to hold the PV and wiring methods in place and to design the wiring methods to allow for any expected movement.

690.31(F) Informational Note explains that expected movement is often included in structural calculations.

690.31(G) Over 1000 Volts DC

New in the 2023 NEC and referred to earlier all the way from **690.7 Voltage** and **690.12 Rapid Shutdown** is the ability to have circuits on a building over 1000V. The reason this is here is because people want to have ground mounted PV systems, but still mount the inverters on the wall of a building.

In order to have dc circuits over 1000V on buildings, we have to comply with all three of the following:

690.31(G)(1) Not permitted for 1- and 2-family dwellings
690.31(G)(2) **Not permitted for buildings with habitable rooms** (Redundancy, since this would include 1- and 2-family dwellings)
690.31(G)(3) The following must apply:

- **Exterior of building**
- **10 feet above grade** (high as a basketball hoop so only those who can dunk can get shocked)
- **Not greater than 33 feet along building** (the length of Sean and Bill's sailboat)

After this **690.31 Methods Permitted** marathon comes the brief …

690.32 Component Interconnections

This section is **specific to building-integrated** (BIPV) **systems.**

Fittings and connectors that are **concealed** at the time of **on-site assembly** and **listed** for such use shall be:

- Permitted for **interconnection of modules**
- Permitted for **interconnection of array components**

Fittings and connectors shall be **at least equal to wiring method** in:

- Insulation
- **Temperature** rise
- **Short-circuit current rating**

These fittings and connectors shall also be able to **withstand the environment** in which they are used.

To repeat, section **690.32 Component Interconnections** was designed specifically for BIPV systems although it does not state this in the NEC, since BIPV (like microinverters) are not defined in the NEC. This is one reason why you need this book, to know these things, otherwise understanding 690.32 blindly would be confusing.

690.33 Mating Connectors

Connectors other than those covered by 690.32 Component Interconnections go here. These connectors are the primary connecters known to solar installers, including your typical **MC4 connector or a microinverter cable connector**.

These connectors **shall comply with all of 690.33(A) through (D)**

The word "mating" was added to the title of this section in 2020. In the 2017 NEC, it was titled "Connectors." This was to indicate what kind of connectors this section is specifying since there are many other types of connectors in the electrical world (e.g., spade connectors).

690.33(A) Configuration

Mating connectors shall be:

- **Polarized** (Example: positive or negative, but not both)
- **Non-interchangeable** (with other receptacles of other systems)

690.33(B) Guarding

Mating connectors shall **guard persons** against inadvertent contact with live parts.

As you can see from playing with a typical MC4 connector, you could not get your finger or tongue into the connector to touch metal. We have done the research, so do not try this at home.

690.33(C) Type

Mating connectors shall be **latching or locking.** ***If*** **readily accessible** and over 30Vdc or 15Vac **then** they shall **require a tool for opening.**

If mating connectors are not of the identical type and brand, then they shall be listed and identified for **intermatability**, as described in the manufacturer's instructions.

Discussion: We are not allowed to use MC4 connectors with Amphenol H4 connectors since they are not listed as "intermatable," although it was a common practice. Here we see there is a pathway for these different connectors to be used together. The process of intermatability is covered in UL 6703. Only two mating connectors currently used in the PV industry are intermatable. They are the Amphenol H4 and the Amphenol UTX. No other combination is officially intermatable. Any salesperson claim that is not backed up by an official UL 6703 certification is simply not true. Incidentally, how can you tell whether a salesman is lying?—their lips are moving.

690.33(D) Interruption of Circuit

Connectors shall be either one of the three following:

690.33(D)(1) Rated for interrupting current
690.33(D)(2) Require a tool for opening and marked either

- "Do Not Disconnect Under Load"
- "Not For Current Interrupting"

690.33(D)(3) Supplied as part of listed equipment

690.33(D)(3) Informational Note: Equipment, such as module level power electronics (including microinverters and dc-to-dc converters) are often listed so that the connectors can be a load-break rated disconnect, even when the connectors say "Do Not Disconnect Under Load" on the PV module connectors themselves.

690.34 Access to Boxes

Junction, pull, and outlet boxes (used for wiring) located behind PV modules shall be installed so that removing the module can make the wiring accessible.

6 Article 690 Part V Grounding and Bonding

Bonding is electrically connecting metal together and grounding is connecting to earth.

690.41 PV System DC Circuit Grounding and Protection

Section 690.41 is only related to PV System DC Circuit Grounding and Protection. Not all PV systems have dc circuit grounding, but all exposed non-current-carrying metal equipment within a PV system that could be energized must be connected to the equipment grounding conductor (690.43).

Here are some simple Article 100 Definitions to get us grounded on grounding:

Grounding Electrode. A conducting object through which a direct connection to earth is established.

Grounding Electrode Conductor (GEC). A conductor used to connect the system grounded conductor or the equipment grounding conductor to a grounding electrode or to a point on the grounding electrode system.

Equipment Grounding Conductor (EGC). The conductive path(s) that provides a ground-fault current path and connects normally non-current-carrying metal parts of equipment together to the system grounded conductor, or to the grounding electrode conductor or both. (Also performs bonding.)

DOI: 10.4324/9781003189862-7

Ground. The earth.

Grounded Conductor. A system or circuit conductor that is intentionally grounded. [Often white or grey and called a neutral.]

Solidly Grounded. Connected to ground without inserting any resistor or impedance device. [An inverter that uses a fuse to detect ground faults is not solidly grounded (on the dc side).]

Functionally Grounded. A system that has an electrical ground reference for operational purposes that is not solidly grounded. [Most PV arrays are functionally grounded; the exception is usually going to be a very small, stand-alone PV system.]

Ground Fault. An unintentional, electrically conductive connection between an ungrounded conductor of an electrical circuit and the normally non-current-carrying conductors, metallic enclosures, metallic raceways, metallic equipment or earth.

Ground Fault Detector-Interrupter (GFDI). A device that provides protection for PV dc circuits by detecting a ground fault and could interrupt the fault path in the dc circuit. [This definition is new in the 2023 NEC, however not new in the PV industry. This is to differentiate between GFCI (below) and detecting and interrupting ground faults on the dc side of a PV system.]

Ground-Fault Circuit Interrupter (GFCI). A device intended for the protection of personnel that functions to de-energize a circuit or portion thereof within an established period of time when a current to ground exceeds the values established for a Class A device. (Class A will trip between 4 and 6 mA.)

Outline of 690.41 PV System DC Circuit Grounding and Protection

690.41(A) PV System DC Circuit Grounding Configurations
  690.41(A)(1) 2-Wire Circuits with One Functionally Grounded Conductor

690.41 PV System DC Circuit Grounding and Protection

Typical ac system grounding is done by electrically connecting a current-carrying conductor to the grounding electrode system in a single place. The grounded conductor is generally colored white or natural gray for solidly grounded systems. Since PV systems often don't produce extreme fault currents, due to the closeness between short-circuit current and operating current, we are functionally grounding rather than solidly grounding dc PV arrays. If you see an older PV system that was installed using the 2014 NEC or earlier, there is often a white dc wire. This is most likely what we call functionally grounded now, but back then, we used to call grounding through a fuse "grounded," which is no longer the case.

Take note that in the 2023 NEC it is clearly indicated that 690.41 is for dc PV circuits, which was always the case, but is more clear now. The old title "system grounding" was confusing to those outside the PV industry, so the title of the section was changed to be more accurate and clear.

690.41(A) PV System DC Circuit Grounding Configurations

Most of the six configurations following are unusual and not worth paying much attention to for most solar professionals. **Paying close**

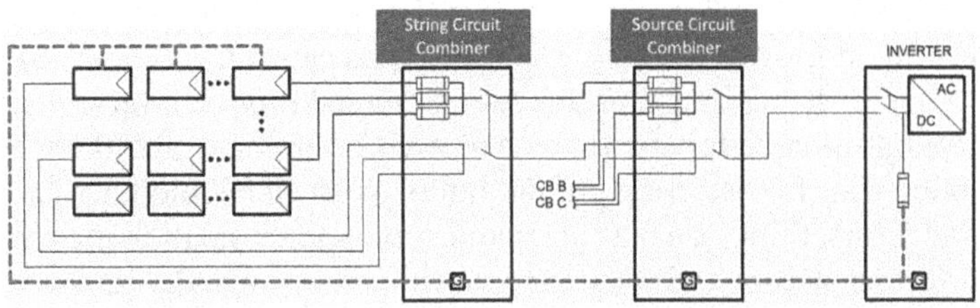

Figure 6.1 Fuse grounded PV array with one functional grounded conductor.

Source: Courtesy Robert Price (2017; modified from Bill Brooks, 2014), www.axissolardesign.com.

attention to 690.41(A)(3) is a good idea, since well over **90% of the inverters installed these days are of the non-isolated type**. For large utility-scale, **MW-scale inverters we most often see the 690.41(A)(1) fuse grounded** (formerly known as grounded) type of inverters, so pay attention to these too. Fuse grounded inverters were the popular inverters of the past for all systems.

690.41(A)(1) 2-Wire Circuits with One Functionally Grounded Conductor

This is our old-style **fuse grounded** solar inverter and was **formerly known as a grounded inverter when applying the 2014 NEC** and earlier. This type of inverter is still common in the large-scale MW inverter realm and seen often when doing maintenance on an old array. This was the most popular inverter in the US until 2013.

The old term often used for this type of inverter's ground-fault detection system is **ground-fault detection and interruption (GFDI); however, as of the 2023 NEC, GFDI applies to all functionally grounded inverters**. This inverter bonds a grounded conductor to the inverter internal grounding busbar through an overcurrent protection device (fuse). When there is a ground fault on the ungrounded conductor, currents will flow through a conductive pathway to the GFDI fuse and open the fuse. The inverter will detect that the grounded conductor is no longer grounded, proceed to turn off the inverter, and then disconnect positive and negative at the inverter to prevent fires (you can still get shocked).

One of the reasons that this fuse grounded type of inverter has gone out of style for less than MW sized inverters, is that there is a blind spot for detecting ground faults. With all PV modules and conductors, things such as glass and conductor insulation are not perfect insulators, and some current will leak through the glass and insulation and complete the circuit through the GFDI fuse. On a small residential system, there is never enough leakage current to cause a problem, but, on larger systems, such as 30kW and up, the milliamps add up, and the current can be great enough to require the inverter to use a larger GFDI fuse. As the fuse gets larger for these leakage currents, it also inhibits the inverter's ability to detect ground faults. Once there is a ground fault that is unseen (blind spot), another ground fault can short-circuit the whole PV array and fires can occur. There are ways that these older systems can be modified by monitoring the current to make them safer, but the

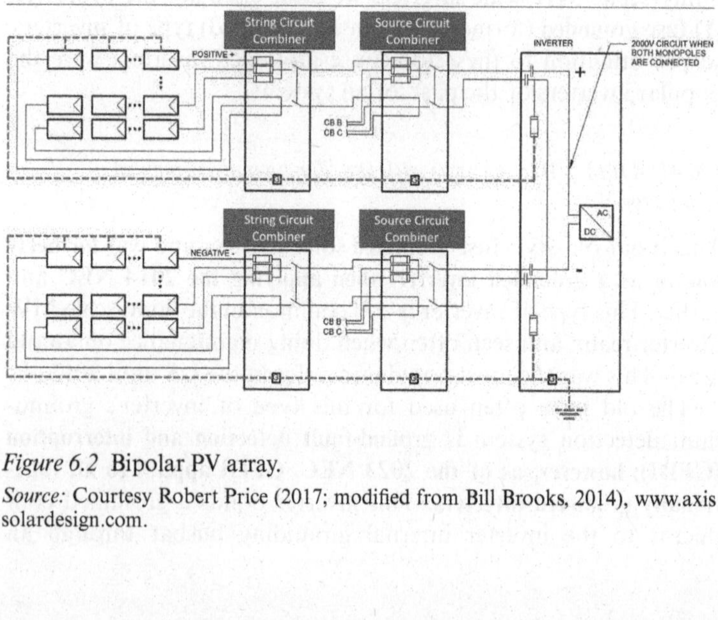

Figure 6.2 Bipolar PV array.

Source: Courtesy Robert Price (2017; modified from Bill Brooks, 2014), www.axissolardesign.com.

easiest thing to do is install a newer, non-isolated inverter, which is cheaper, safer, and more efficient.

690.41(A)(2) Bipolar Circuits with a Functionally Grounded Center Tap

We have mentioned bipolar arrays a number of times in this book, and you can look in the index to read about bipolar arrays at every mention. Bipolar arrays are usually only seen in large utility-scale PV systems. We have not seen many between 2012 and 2023 being installed, but they may be making a comeback in some cases before the 2023 NEC goes out of style, according to our resident futurist.

690.41(A)(2) applies to these bipolar arrays with a functional ground reference (center tap).

A few people think of bipolar arrays as a way to sneak around the requirement for maximum voltage and double the system voltage. Bipolar arrays will give you double voltage benefits, so there is truth; however, now that we have ground-mounted PV arrays of 1500V to ground, there is less of an incentive to double the voltage, since 1500V is a lot and jumping to 3000V for efficiency gains may not be worth doing.

690.41(A)(3) Arrays Not Isolated from the Grounded Inverter Output Circuit

This is the **most popular inverter today**, representing over 90% of inverters installed. **Formerly and commonly (though somewhat inaccurately) known as an "ungrounded inverter"** in the 2014 NEC, this inverter is now called a **functionally grounded inverter** and, according to 690.41(A)(3) text, we can officially call this inverter a **not-isolated** inverter or, as we prefer, a **non-isolated inverter**.

In Figure 6.3 on the next page Vin would be input voltage to the inverter. Vs is the source voltage (center-tap split-phase ac transformer).

The **non-isolated** gets its name because the inverter is not isolated from the grounded ac transformer. The reason that it was formerly thought of as ungrounded is because the PV array is ungrounded when the system is offline and grounded when operating. When this inverter is operating at 400V, we could expect the positive conductor to measure 200V to ground and the negative to measure −200V to ground. This array is referenced to ground and does not

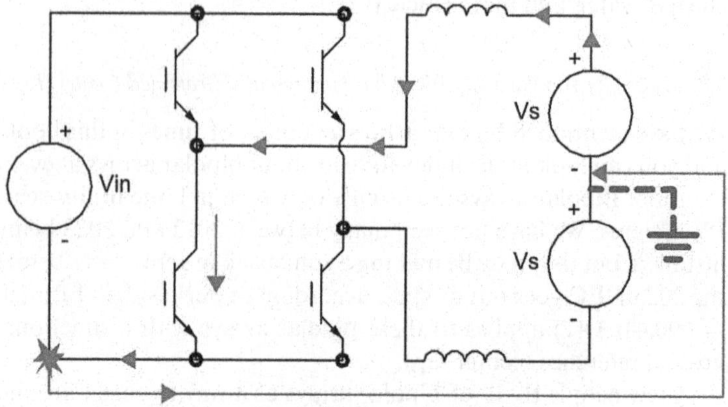

Figure 6.3 Non-isolated inverter showing ground-fault pathway.
Source: Courtesy Robert Price (2017; modified from Bill Brooks, 2014), www.axissolardesign.com.

have a voltage that is randomly floating around or isolated by a transformer when the inverter is operating.
These inverters are:

- cheaper
- safer
- lighter
- more efficient

than your old style, 690.41(A)(1) 2-wire PV arrays with one functionally grounded conductor, inverters (fuse grounded).

Cheaper because no transformer to manufacture.
Safer because more sensitive ground-fault protection.
Lighter and more efficient since no isolation transformer.

Non-isolated inverter ground-fault detection:

Non-isolated inverters can detect ground faults much more sensitively than fuse grounded inverters.

Fuse grounded inverters have to have an allowance for leakage currents and the fuse must be upsized as more PV is added to the system.

Non-isolated inverters have different methods of detecting dc ground faults, such as:

- Insulation testing
  - As the inverter wakes up in the morning, a quick pulse of voltage will be sent out along the current-carrying conductors into the array. If there is lower than expected resistance to ground, then there is a failed insulation test and signs of a ground fault. The inverter will not be allowed to start. We want to see millions of ohms of resistance in the insulation of a conductor and hundreds of thousands of ohms of resistance in a string of modules.
- Comparing positive and negative currents (residual current monitor)
  - While operating, if positive does not equal negative, then the electrons must be taking an alternate path, otherwise known as a ground fault (or Martians are feeding on the electrons).

> Using non-isolated inverters prevents fires since it can be up to 3000 times more sensitive to ground faults than old-school fuse grounded inverters! That's as sensitive as the people who are offended by our dad jokes.

690.41(A)(4) Ungrounded PV Arrays

The NEC definition of an ungrounded array (figure 6.4 next page) is different from the 2014 and earlier editions of the NEC. This truly ungrounded PV array is **not** the 690.41(A)(3) *"now known as non-isolated array" (formerly known as ungrounded)* but is an array that has no functional reference to ground. Imagine a fuse grounded inverter and then you threw the fuse away and the inverter was programmed to still work. This is essentially what we are talking about.

Now the only way to have an ungrounded array on an inverter connected to a grounded utility transformer on the ac side, is where the inverter has an isolation transformer, but with no fuse

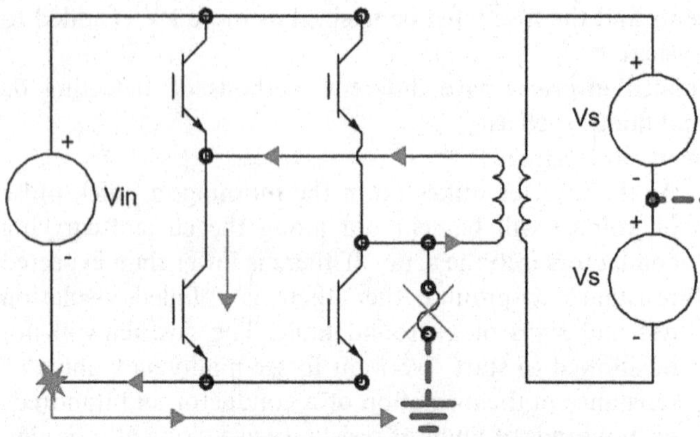

Figure 6.4 Ungrounded PV array a.k.a. transformer-isolated inverter. *Source:* Courtesy Robert Price (2017; modified from Bill Brooks, 2014), www.axissolardesign.com.

or connection between a current-carrying conductor and ground as in Figure 6.4. This array would be considered floating since the voltage could change with reference to ground. This inverter is uncommon and only currently used with some large utility-scale inverters. The array is monitored with a sensitive insulation tester on a continuous basis. This is a very simple and effective method of ground-fault detection.

There is no way to look at an inverter and know whether it is an ungrounded inverter. We are unaware of any ungrounded inverters currently listed in the United States. All the ungrounded units currently running in the US are in large-scale plants and have been certified to IEC (International Electrotechnical Commission) standards.

690.41(A)(5) Solidly Grounded PV Arrays

A solidly grounded PV array (figure 6.5 next page) is one where there is a connection between a current-carrying conductor and the grounding electrode system that is not a fuse and is solid! This is like your house's ac wiring.

Many old-style stand-alone systems, as well as some more modern stand-alone systems, are solidly grounded. Many dc PV systems are also solidly grounded.

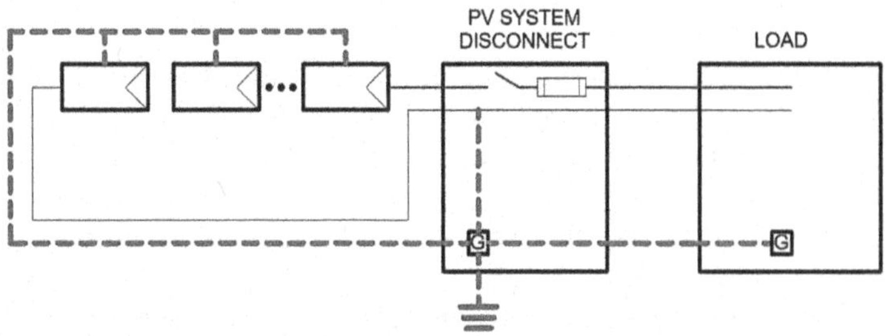

Figure 6.5 Solidly grounded PV array.

Source: Courtesy Robert Price (2017; modified from Bill Brooks, 2014), www.axissolardesign.com.

Coming up soon in 690.41(B) [formerly 690.41(B) exception], we will mention that if an array is solidly grounded, has one or two PV source circuits, and is not on a building, then ground-fault protection is not required. This way we can solidly ground the system.

As we can see in Figure 6.5, the grounding busbar is connected solidly to a grounded conductor. This is also an example of a dc direct-coupled PV system, which is common for water pumping systems in the California foothills.

Solidly Grounded Systems are Very Common, Just Not for PV Arrays

We are used to solidly grounded systems with typical electrical systems. The reason it is uncommon with PV systems is because of the lack of high short-circuit currents, due to the current-limited characteristics of PV. With low short-circuit currents, we cannot rely on standard overcurrent devices to clear faults. We can only detect faults and shut down the system as a result of the fault. With a wild PV circuit, everything including the overcurrent protection is sized based on at least 156% of a short-circuit current. So with a short it will not open a fuse.

690.41(A)(6) Circuits Protected by Equipment Listed and Identified for Use

PV technology is fast-moving, and the Code has left open a way for newer inventions to be Code-compliant if they are listed and approved. Over the years, several inverters have been certified to this option. One key example in somewhat recent history was the Advanced Energy Inverters that had an array configured in a bipolar-ish fashion.

These larger inverters were configured similarly to bipolar PV systems but had several features that were very different from bipolar arrays. Those inverters were among the safest inverters back when they were manufactured. 690.41(A)(6) makes sure the NEC does not disrupt inverter innovation for your next genius inverter idea. Please cut us in.

690.41(B) DC Ground-Fault Detector-Interrupter (GFDI) Protection

Circuits over 30V or 8A (most PV modules) shall have GFDI as seen in 690.41(B)(1) and 690.41(B)(2) below.
Solidly grounded PV source circuits are only allowed without GFDI if:

- No more than 2 in parallel
- Not on or in buildings

(This was formerly an exception in the 2017 NEC.)

Discussion: It may be less clear now, that the part about solidly grounded systems with 2 or less in parallel, and not on or in buildings, is no longer an exception. To be clear, you can still have solidly grounded systems without GFDI that are over 30V and 8A. The organization has changed, but the implementation has stayed the same, as with most NEC changes.

690.41(B) Informational Note

Not all equipment that dc PV arrays are connected to has GFDI. **If GFDI is not included**, then in the installation manual there is often a **warning statement that indicates GFDI is not included**.

Equipment without GFDI is often smaller stand-alone equipment and not interactive inverters.

Discussion: The dc ground fault protection has to do with what you are connecting your PV array to, so it is not for a battery inverter in this Article 690 PV context. It is usually for an interactive inverter PV input or a charge controller PV input.

690.41(B)(1) Ground-Fault Detection

Ground-fault detection must meet all the following requirements:

- Detect ground faults in PV array dc circuits
- Detect ground faults in any functionally grounded conductors
- Listed for PV GFDI protection

Dc-to-dc Converters: Dc-to-dc converters sometimes are not listed to provide ground-fault protection and must work with listed GFDI protection equipment.

Informational Note: Some dc-to-dc converters without ground-fault protection on their source (PV) side may prevent other ground-fault protection equipment from working properly. **In other words, in some dc-to-dc converter cases:**

> GFDI + GFDI = No GFDI (two rights can make a wrong—sounds political)

690.41(B)(2) Faulted Circuits

Faulted circuits shall be **controlled** by **either one** of the following of 690.41(B)(2)(1) or 690.41(B)(2)(2):

- **690.41(B)(2)(1) Current-carrying conductors auto-disconnect**. Current-carrying faulted conductors of the circuit will automatically disconnect.
- **690.41(B)(2)(2) Power off, interrupt faulted PV dc circuits** from the ground reference in **functionally grounded systems**.

Inverter or charge controller fed by faulted circuit will **automatically do both**:

- Stop supplying power to output circuits
- Interrupt PV system dc circuits from ground reference in functionally grounded system

Discussion: A properly **UL 1741 listed inverter** will have been tested and listed for ground-fault protection.

Equipment Certification, UL and ANSI

Equipment certification is a de facto requirement of the NEC because it requires all equipment to be evaluated for the intended use in Section 90.7 Examination of Equipment for Safety. Underwriters Laboratories (UL) has been tasked by the ANSI (American National Standards Institute) to establish certification standards for the safety of PV equipment. There are numerous UL standards that are used for PV equipment. UL 1741 is the standard used for certifying inverters and other PV related electronic equipment, such as charge controllers and dc-to-dc converters.

690.41(B)(3) Indication of Faults

GFDI equipment must indicate that there was a ground fault at a readily accessible location.

690.41(B)(3) INFORMATIONAL NOTE

Examples of indication methods include:

- Remote indicator light
- Display monitor
- Signal to monitored alarm system
- Notification by web-based services

Discussion: Web-based services can be a text message or an app notification. (Notice that smoke signals were not included, since that could be mistaken that the building is on fire—maybe it is if we did not address the fault…)

690.42 Point of PV System DC Circuit Grounding Connection

690.42(A) Circuits with GFDI Protection

Circuits protected by GFDI equipment according to 690.41(B) DC Ground-Fault Detector-Interrupter (GFDI) Protection, shall have circuit-to-ground connection made via GFDI equipment.

690.42(B) Solidly Grounded Circuits

Solidly grounded systems are grounded from any single point of PV dc system to the grounding electrode system in 690.47(A) Building or Structure Supporting a PV System (690.47 Grounding Electrode System is coming up on page 147).

Ground Fault Protection, GFDI and GFCI

Ground-fault protection is, in essence, determining if there is a connection to ground by a current-carrying conductor that was not meant to be. To determine if there is an extra connection (ground fault), then a ground-fault protection device must be keeping tabs on the state of the current-carrying conductor vs. ground.

In any electrical system, more than one point of system grounding is called a ground fault. If there were two points of system grounding, then there would be a parallel pathway for current to flow along the grounding system or through equipment.

In the 2023 NEC the GFDI term was put into the NEC to differentiate between ground fault circuit interruption (GFCI) and ground-fault detection and interruption (GFDI). The term GFDI has been around for a while in the industry and often associated with fuse grounded systems. However, it is clear now that GFDI includes all PV dc ground fault protection and has nothing to do with GFCIs.

Outline of 690.43 Equipment Grounding and Bonding

690.43(A) Photovoltaic Module Mounting Systems and Devices
690.43(B) Equipment Secured to Grounded Metal Supports
690.43(C) Location
690.43(D) Bonding Over 250V

Grounding shall comply with **250.134 Equipment Fastened in Place or Connected by Permanent Wiring Methods (Fixed)** and **250.136 Equipment Secured to a Metal Rack or Structure**. 250.136 sends you back to 250.134, which sends you to other places that send you to other places, as is common for the very huge **Article 250 Grounding and Bonding**.

Equipment grounding of PV systems is theoretically the same as equipment grounding of other systems; hence we refer to and take the rules from Article 250 when applying the Code to our PV system grounding systems.

For instance, 690.43 directs us to 250.134, which directs us to 250.32, which directs us to 250.118 and so on and so forth.

Why Inspectors Like to Call You on Grounding and Bonding

Since equipment grounding of a PV system is similar to and takes much of its requirements from Article 250 Bonding and Grounding, inspectors who are not as familiar with PV systems as they are with other systems will often put a lot of effort into inspecting what they know, which is how to inspect grounding and bonding.

Here are some of the places that 250.134 can take you:

250.134 Equipment Fastened in Place or Connected by Permanent Wiring Methods (Fixed)

Here we are referring to bonding (grounding) non-current-carrying metal parts:

- You can be connected to the grounded circuit conductor as permitted by either:

- o 250.32 Buildings or Structures Supplied by a Feeder(s) or Branch Circuit(s)
- o 250.140 Frames of Ranges and Clothes Dryers (690 brings you here, Wow!)
- o 250.142 Use of Grounded Circuit Conductor for Grounding Equipment

- Or you can be connected by one of the following:
  - o 250.134(1) by connecting to equipment grounding conductors as permitted by 250.118(2) through (14)
    - There is a **typo in the 2023 NEC** here, since 250.118 was reorganized, so it should say 250.118(A)(2) through (14). Perhaps by the time you read this, the typo will have been un-typoed.
    - 250.118(A) lists **types of EGCs permitted**
    - 250.118(A) sends us to enough places to write a book on grounding. If you are a **super-grounding-nerd**, you can read **Mike Holt's book on "Bonding versus Grounding,"** and we will be impressed when you stump us, and we expertly dodge your question in front of hundreds of people at a big important conference.
    - Note: Soares is and has been wrong when it used the term "grounding."
  - 250.134(2) option is by connecting to an **EGC that is contained within the same raceway, cable or run with circuit conductors.**
    - 250.134(2) Exception 1: As provided in 250.130(C) Replacement of Nongrounding Receptacle or Snap Switch and Branch Circuit Extensions, EGC shall be allowed to run separately.
    - **250.134(2) Exception 2: For dc circuits, EGC shall be permitted to run separately. (This is relevant to us dc people!)**

Why Equipment Grounding?

A good reason for equipment grounding is to protect people. If a hot wire got loose and touched a piece of metal that was not connected to the equipment grounding conductor, then someone could touch that piece of metal and get shocked. Another good reason for grounding is a pathway for ground-fault protection to signal the inverter when there is a ground fault. Without an equipment grounding conductor, ground-fault protection would not work properly.

690.43(A) Photovoltaic Module Mounting Systems and Devices

Equipment that is listed, labeled, and identified for bonding PV modules (frames) is permitted for equipment grounding and bonding of PV modules. (Also, this equipment may be used to bond adjacent modules to each other.)

690.43(A) Informational Note: UL 2703 and UL 3703 are the standards used for bonding PV module frames. **UL 3703 is used for trackers.**

UL 2703 Standard: Standard for Mounting Systems, Mounting Devices, Clamping/Retention Devices, and Ground Lugs for Use with Flat-Plate Photovoltaic Modules and Panels.

When we think of UL 2703 listed racking systems, we often think of a clamping device that holds a PV module or piece of equipment in place and has a sharp point that will pierce the anodized coating of a piece of aluminum and make a Code-compliant equipment grounding pathway.

A big benefit to PV installers for using UL 2703 listed racking is the reduction in time and materials required by not having to undertake time-consuming grounding methods of the past, such as using lugs on each PV module with a collection of washers, antioxidant chemicals and using emery cloth to rub off the anodized coating of a module frame to get a good connection.

690.43(B) Equipment Secured to Grounded Metal Supports

Grounded metal supports are support structures that PV modules are mounted on that are already connected to the earth by any means permitted in Article 250. In order to bond PV equipment to grounded metal supports, the grounding devices must be:

- Listed = On a National Recognized Testing Lab (NRTL) list (search "OSHA NRTL" to find the list).
- Labeled = Has a label from the NRTL
- Identified = For specific use (no off-label usage)

Metallic support structure sections shall be **either**:

- Connected via identified bonding jumpers
- Connected to equipment grounding conductor (EGC) and support structure shall be identified

690.43(C) Location

- EGCs shall be permitted to run separately from the PV system conductors **within the array**
- If the EGC is run separately, follow **250.134 Equipment Fastened in Place or Connected by Permanent Wiring Methods (Fixed)**, which we just covered (see page 140)

Discussion: **690.43(C) in the 2020 NEC was called "With Circuit Conductors"** and required us to run the EGC with circuit conductors, and now in the 2023 NEC we are more liberal and allow our EGC to be run separate from the circuit conductors, like other dc systems.

690.43(D) Bonding Over 250 Volts

For **solidly grounded** systems over 250V (rare), **250.97 Bonding for Over 250 Volts to Ground** shall apply.

Here we see requirements about continuity of metal raceways and metal sheathed cables for solidly grounded systems over 250V to ground.

Where concentric or eccentric knockouts are encountered, there may not be sufficient continuity to clear large circuit breakers in a fault. This is more commonly referenced for ac systems over 250V to ground, since solidly grounded PV systems, especially those over 250V to ground, are rare.

No Bonding Bushings Required!

In the 2017 NEC and earlier, this 250.97 requirement for concentric and eccentric knockouts was not only for solidly grounded systems, and many inspectors are still enforcing the older requirements and making you install bonding bushings on functionally grounded dc PV systems over 250V. Politely show your inspector 690.43(D) if you do not like to spend time and money on bonding bushings where they are not required.

Sensitive ground-fault equipment required in functionally grounded systems can effectively use concentric and eccentric box connections without failing in a fault and therefore doesn't need to follow 250.97.

Table 6.1 NEC Table 250.122 EGC based on OCPD

Rating or setting of automatic overcurrent device in circuit ahead of equipment, conduit, etc., not exceeding (amperes)	*Size (AWG)*	
	Copper	*Aluminum*
Aluminum or copper-clad aluminum		
15	14	12
20	12	10
60	10	8
100	8	6

Source: Courtesy NFPA.

690.45 Size of Equipment Grounding Conductors

PV system circuit equipment grounding conductors are sized in accordance with **Table 250.122 Minimum Size of Equipment Grounding Conductors**, which bases the EGC off the size of the overcurrent protection device.

If there is no overcurrent protection device (such as with one or two PV source circuits) then the value calculated in **690.9(B) Overcurrent Device Ratings (Isc × 1.56 for "wild PV")** for an assumed overcurrent protection device will be used in place of an overcurrent protective device value.

Recall that 690.9(B) (covered on page 57) calculates an overcurrent protection device based on either:

- 125% of maximum current calculated in 690.8(A)
    - Maximum circuit current × 1.25
    - Isc × 1.56 for PV source circuits (since Isc × 1.25 = maximum circuit current for "wild PV')
- Maximum current and application of adjustment and correction factors (see Chapter 12 Wire Sizing)
- Electronic overcurrent protection device listed to prevent backfeed on PV side of device
- Equipment grounding conductor size shall not be smaller than 14 AWG

Discussion: It is interesting to note that, in many cases, multiple PV output circuits are not required to have more than a single 14 AWG equipment grounding conductor when protected in a raceway. Many solar installers will size the equipment grounding conductor to be the same size as the current-carrying conductor for conductors up to 10 AWG. This is often done to decrease the possibility of a well-intentioned inspector questioning a Code-compliant 14 AWG equipment grounding conductor.

Just to make a point here, it would be Code-compliant to have 8 PV source circuits come off a rooftop in a raceway with 16 different 10 AWG current-carrying conductors along with a single 14 AWG equipment grounding conductor.

Since the smallest copper equipment grounding conductor in Table 250.122 is a 14 AWG, which is sized based on a 15A overcurrent device, then we can calculate the highest short-circuit current value that will fit on a 15A fuse and use a 14 AWG equipment grounding conductor as follows:

Isc × 1.56 = 15A
Isc = 15A/1.56 = 9.6A

Therefore, if a PV module has an Isc value of over 9.6A, then it would need greater than a 15A fuse and greater than a 14 AWG equipment grounding conductor. And a PV module with an Isc over 9.6A would have a maximum series fuse rating of 20A (unless you could find a future module that had an Isc greater than 20A/1.56 = 12.8A).

690.45 specifically states that "we do not have to increase EGC size for voltage drop considerations on either the dc or ac side of a PV system." Since PV is the opposite of a load, current does not increase because of voltage drop. There are no NEC or safety requirements for PV systems regarding voltage drop. A 50% voltage drop would be illogical and very wasteful, but not necessarily a Code violation. This contradicts and supersedes Article 250; just because Article 250 penalizes you for doing the right thing (oversizing your current-carrying conductors) does not mean that 690 is going to go along with this reverse logic. Oversizing a current-carrying conductor makes your PV system safer and more efficient, regardless of whether you up-size your EGC. Remember NEC Chapters 5 through 7 modify and supersede Chapters 1 through 4.

Exposed to Physical Damage = 6 AWG Minimum Bare Copper

Oftentimes the equipment grounding conductors (EGCs) under the array are considered to be **exposed to physical damage** and, if that were the case, the bare copper equipment grounding conductor would need to be a **6 AWG** size minimum. Oftentimes in places with significant weather, such as snow and ice, the AHJ will require a 6 AWG bare copper equipment grounding conductor if exposed bare copper is used. In the Deep West, where the wind is relatively mild, oftentimes you see 10 AWG EGCs under the array, but in the East, you often see the AHJ require 6 AWG bare copper and some solar installers will use smaller green USE-2 to conserve copper.

If you are trying to talk your inspector into agreeing that you do not require a 6 AWG EGC, you can reference 250.120(C) Equipment Grounding Conductors Smaller than 6 AWG, where it says that you need to be protected from physical damage in a raceway or cable armor, unless it is **installed within** the hollow spaces of framing members of buildings or **structures** and where it is not subject to physical damage. Since a PV array is a structure, you would have a good point. We are not saying that you would win the debate, but you would have an intelligent point to make. The unanswered question is, "What is considered subject to physical damage?" and the answer is regional and subjective. In the end, how much do you really want to argue with an inspector when you are trying to get your project finalized?

It is interesting to note that, in the Chicago area, according to at least 10 people whose jobs it is to meet with inspectors every day, they require more heavy-duty intermediate metal conduit (IMC) or rigid metal conduit (RMC) instead of EMT. This is due to the extreme weather, but they allow 10 AWG EGCs under the array (because Bill told them it was okay 20 years ago).

In other locations, installers are allowed, as they should be, to use green 10 AWG USE-2 wire under the array.

690.47 Grounding Electrode System

Are you one of the top PV experts in the country? Lace up your boxing gloves and welcome to 690.47, the section of 690 where there are heated debates on how to interpret how grounding electrode systems are meant to be interpreted in the Code. Try not to take it personally if you have a different idea than your colleagues. Good debate can be fun! Especially when you are right and everyone else is wrong. YouTube search "2014 NEC 690.47(D) Mike Holt" for some excitement, where you will see Bill and Sean along with Mike, who endorses this book. Endorsement to your millions of NEC books, back at you Mike!

There are many factors to deal with here and grounding electrode systems outside of PV are also hard to understand because the way they are implemented differs regionally. Factors such as wet earth, dry earth, lightning vortexes, corrosion, living on a rock, stray currents, proximity to large power production systems and more will influence people's opinions on how to properly connect to earth.

Refresher:

EGC = equipment grounding conductor
GEC = grounding electrode conductor

Figure 6.6 Mike Holt 2014 PV NEC team.

Outline of 690.47 Grounding Electrode System

690.47(A) Buildings or Structures Supporting a PV System
- 690.47(A)(1) ECG Only Connection to Ground If Not Solidly Grounded
- 690.47(A)(2) Solidly Grounded Systems Dc Grounding Electrode Conductor (GEC)

690.47(B) Grounding Electrodes and Grounding Electrode Conductors

690.47 Grounding Electrode System was once again changed in the 2023 NEC, with a bit less pointing to Article 250 in 690.47(A). In the 2020 NEC, 690.47(A) broadly pointed to 250 **Part III Grounding Electrode System and Grounding Electrode Conductor**, which is no longer the case.

690.47(A) Buildings or Structures Supporting a PV System

690.47(A) says that buildings or structures supporting a PV system shall utilize a grounding electrode system installed in accordance with **690.47(B) Grounding Electrodes and Grounding Electrode Conductors.**

Additionally, 690.47(A) tells us that PV array equipment grounding conductors (EGC) shall be connected to a grounding electrode system in accordance with:

- **Article 250 Part VII Methods of Equipment Grounding Conductor Connections**
- Connection is in addition to **690.43(C) Location,** which tells us now that the EGCs are permitted to be run separately from the PV system conductors within the array (see page 143).

Also building and structure array EGCs shall be sized in accordance with **690.45 Size of Equipment Grounding Conductors** (see page 144), which directs us to **Table 250.122 Minimum Size of Equipment Grounding Conductors** (see page 146).

Back to where we were. One of the two following will apply. **690.47(A)(1) is for systems that are not solidly grounded**, which is the majority of everything, and 690.47(A)(2) will apply to the extremely rare solidly grounded PV system not on a building.

690.47(A)(1) Not-Solidly Grounded Systems on buildings are allowed to have the EGC be the only connection to ground. This definitely means that **for most systems installed today, we do not need a dc GEC like the old days**. Dc grounding electrode conductors are just out of style.

690.47(A)(2) Solidly Grounded Systems on structures would have to comply with 690.41(A)(5) on page 134 and we size the dc GEC based on Table 250.166 Size of Direct-Current Grounding Electrode Conductor. Remember, these systems are only allowed where not on buildings so this is **rare** and we will not go too deep and waste your extremely valuable time.

690.47(B) Grounding Electrodes and Grounding Electrode Conductors

A grounding electrode system includes pieces of metal connected to earth and other pieces of metal connecting those pieces together and then to the grounding electrode system via a grounding electrode conductor (GEC).

690.47(B) has been the controversial "back and forth" over the years. The conflict was this: should we have an additional auxiliary electrode (or electrodes) or not? In the 2008–2014 NECs, it was often interpreted as required, which upset people, but many did not install them. In the 2017 NEC, this section was reworded to make it clear that the additional auxiliary electrode was optional. Once again, in the 2020 and 2023 NEC it is optional, but no longer called auxiliary. Now it is just termed **additional electrode(s)**.

Auxiliary Electrodes are electrodes that are **attached to the equipment grounding system** and do not follow the rules of typical electrodes. Now, apparently, we have to follow the typical rules for something optional.

In the **2008 to 2014 NEC** 690.47(D), Additional Auxiliary Electrodes for Array Grounding said that these electrodes **shall** be installed with two exceptions. These two exceptions essentially made it not required for systems on buildings. The exceptions were so poorly worded that few code enforcers allowed the exceptions and the result was that electrodes got installed on buildings with perfectly acceptable grounding electrode systems. This is now a low-key **optional** thing that someone can do if they want. One of the reasons for this being controversial was because lightning strikes to the earth can send a **wave of voltage**. If the wave hits an electrode and causes different voltages at different electrodes that are attached through equipment, then the equipment can sizzle and pop.

What these optional electrodes are: If you have an array on your roof, you are permitted to take a conductor from the array straight to a ground rod. This would also go for a ground mount; however, one way you can look at it is if your array is put in concrete or earth with metal, every post is electrically an additional electrode. You can connect the 690.47(B) electrodes to the PV arrays on buildings to the metal structural frame of the building if you want. Keep in mind that this connection does not substitute for the EGC connection to the inverter. Without that EGC connection at the inverter, the ground-fault detectors will not work properly.

Get grounded! At least functionally.

7 Article 690 Part VI Source Connections + Marking and Labeling Summary

For Marking and Labeling Summary, go to page 155.

In the 2023 NEC, there was a lot of rearrangement towards the end of Article 690, and the **2020 NEC version of this chapter in *PV and the NEC* covered**:

2020 NEC 690 Part VI Marking
2020 NEC 690 Part VII Connection to Other Sources
2020 NEC 690 Part VIII Energy Storage Systems
2020 NEC Other Material no longer in 690 as of 2017 NEC

Bill and his Code Making Panel 4 friends were busy **consolidating and rearranging for the 2023 NEC** and now we are left with just one short Part VI to finish off 690:

2023 NEC 690 Part VI Source Connections

This reduction of parts does not mean that all the other requirements that were in this chapter are gone. They just moved things and got rid of redundancies and irrelevancies. For instance, marking requirements for the rapid shutdown sign were moved to 690.12 Rapid Shutdown. Some of the marking requirements disappeared since they were not saving lives and property, which the NEC is for. Other marking requirements did not need to be in the NEC, since things like the ac PV module specifications are already required to be on the backs of the ac PV modules according to the UL listing.

What we have decided to do to keep this chapter long enough to be a chapter and to help you have a single place to go for all

DOI: 10.4324/9781003189862-8

your marking and labeling requirements, is to include marking and labeling in this chapter beginning on page 155. Many of these requirements are in other places in this book; however, people like having a single place to go for all these requirements.

Outline of 690 Part VI Source Connections

(In the 2020 NEC we had Part IV Marking and Part VII Connections to Other Sources.)

690.56 Identification of Power Sources (formerly 690.56 in Part VI Marking)
690.59 Connection to Other Sources (formerly 690.59 in Part VII Connection to Other Sources)
690.72 Self-Regulated PV Charge Control (formerly in Part VII Energy Storage Systems)
- 690.72(1) PV source circuit matches current and voltage of battery cells
- 690.72(2) 1hr < 3% charge requirement

690.56 Identification of Power Sources

In the 2023 NEC, all 690.56 does is send you to 705.10 Identification of Power Sources. To abbreviate here: Power sources need labeling to show location and to say, "CAUTION MULTIPLE POWER SOURCES." A new 2023 NEC thing is we need to leave a phone number. For details go to page 195.

(OMG, what's the number for 911?!—Sorry, blonde runs in the family.)

With PV systems, energy storage systems, and the grid, we can have power pushing in all directions, therefore it is prudent to identify what is going on for those who will be puzzled in the future. As the future happens, we will see more things connected, and with energy management systems (a.k.a. power control systems) we will discover that more things will be networked to work in harmony for a better grid experience.

690.59 Connection to Other Sources

690.59 Connection to Other Sources is very brief and sends us to **Article 705 Parts I and II**

- 705 Part I General (most of 705)
- 705 Part II Microgrid Systems (about 1 page of 705 to end)

Note that Article 712 Direct-Current Microgrids was removed in the 2023 NEC and is included in 705 Part II.

The majority of Article 705, including the interconnection requirements in 705.11, 705.12, and 705.13, is in Part I of 705. See page 196.

690.72 Self-Regulated PV Charge Control

Self-regulated PV systems have the benefit of not needing a charge controller but have the drawback of not being able to work at the optimal part of the IV curve. These systems are often designed so that they will operate towards the Voc end of the IV curve, so that an overcharge will not occur, or they can be designed so that the PV is undersized relative to the battery, yet so is the load. With the advent of less expensive and more reliable PV electronics, self-regulating systems are less common. An example often used for a self-regulating system is a buoy in the ocean with 35 solar cells, charging a 12V battery to operate a flashing light that takes very little energy. Another example is my small solar module that used to keep my old internal combustion engine (ICE) car battery topped off when I went away for extended periods of time. No charge controller is needed.

690.72 says: "A PV source circuit shall be considered to comply with the requirements for charge control of a battery without the use of separate charge control equipment if the circuit meets both of the following" which is 690.72(1) and 690.72(2) below.

690.72(1) Matching PV to Battery

The voltage of the PV should match the voltage of the battery(s) paying attention to the IV curve.

690.72(2) No Greater Than 3% per Hour

The current of the PV should not overcharge the battery(s) and we draw the line at no greater than 3% per hour of battery capacity, or as recommended by battery manufacturer.

Self-Regulating Systems

With self-regulating PV charge control, it is like we have a PV source circuit that is a silicon-based charge controller, using silicon to control things, when it is not a computer chip!

There is rumored to be a NABCEP PV Associate (PVA) exam question on self-regulating charge control, which is a nuisance for a PVA exam prepper like myself, since I have a question every week about self-regulating systems, and you never see them in real life, only on the PVA exam.

Self-regulating systems are designed for batteries that can take a float charge and lithium-ion batteries cannot take a float charge, since they are very efficient and do not self-discharge much at all. Sean recommends noting this in the NEC or removing 690.72, since if someone designed a self-regulating system to charge 3% of battery capacity in an hour with lithium-ion batteries, it could be a fire hazard. Lithium-ion batteries need a battery management system (BMS) to manage the charge, so they would not be self-regulating with a BMS by definition.

The idea of PV self-regulation is still creative and very neat!

We do not recommend self-regulating charge control of lithium batteries on Boeing 787s. Sky fire is not cool.

This is the End of 690!

Marking and Labeling Summary

Marking and labeling of renewable energy systems can sometimes take you on a long trip throughout different chapters of the NEC. We are going to sum it up here, at least for the more common marking and labeling requirements. Take note that the labeling requirements of the 2023 NEC were reduced from those of the 2020 NEC, so if you think you are missing something that was in the earlier versions, this can be the reason. For the rest of this chapter, we are going to have a license to use the word labeling for marking, placards, etc.

We will skip Article 480, since 690 tells us to use Article 706, plus many of the 480 requirements are the same as in 706.

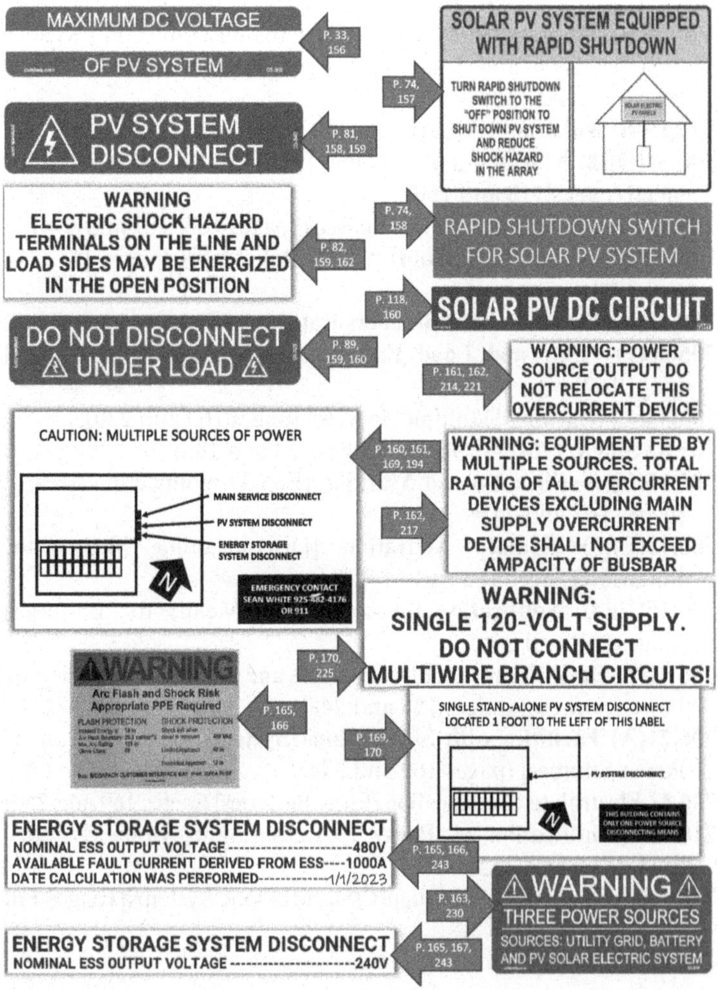

Figure 7.1 Shows the label image and the page number.

List of Labeling Requirements

690.7(D) Marking DC PV Circuits (page 33 and this page below)
690.12(D) Buildings with Rapid Shutdown (pages 73 and 157)
690.12(D)(1) Buildings with More than One Rapid Shutdown Type (pages 75 and 158)
690.12(D)(2) Rapid Shutdown Switch (pages 76 and 158)
690.13(B) Marking [Photovoltaic Systems Disconnecting Means, includes Line and Load Energized in Open Position label] (pages 81 and 158)
690.15(B) Isolating Device (pages 118 and 160)
690.31(D)(2) Marking and Labeling [SOLAR PV DC CIRCUIT label] (pages 118 and 160)
705.10 Identification of Power Sources (pages 160, 169 and 194)
705.12(B)(2) and 705.12(B)(4) "120% Rule" label (pages 161, 214 and 221)
705.12(B)(3) "Sum of the Breakers Rule" label (pages 162 and 217)
705.20(7) "Line and Load Energized in Open Position" Label (page 162)
705.30(C) Marking [Multiple Sources] (pages 163 and 230)
705.82 Single 120-Volt Supply (pages 163 and 188)
706.15(C) Notification and Marking [ESS Disconnecting Means] (pages 165 and 243)
706.15(E)(3) Remote Activation [Disconnecting Means for Batteries] (pages 168 and 245)
706.15(E)(4) Notification [Disconnecting Means for Batteries] (pages 168 and 245)
706.21(A) Facilities with Utility Services and ESS [Identification of Power Sources] (pages 168 and 248)
706.21(A) Facilities with Stand-Alone Systems [Identification of Power Sources] (pages 169 and 248)
706.41 Electrolyte Classification [Flow Batteries] (pages 169 and 250)
710.10 Identification of Power Sources [Stand-Alone Systems] (pages 169 and 252)
710.15(C) Single 120-Volt Supply [Stand-Alone Systems] (pages 170 and 255)

Marking and Labeling Details

690.7(D) Marking DC PV Circuits [in 690.7 Maximum Voltage] (also page 33 of this book)

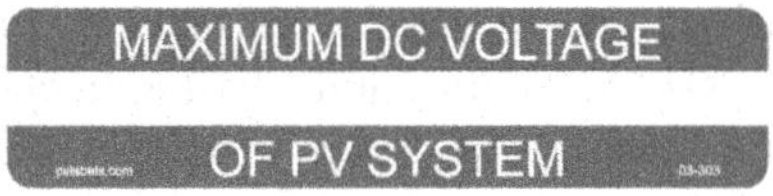

Figure 7.2 690.7(D) Max DC voltage label.
Source: Courtesy pvlabels.com.

Requirement for marking highest maximum dc voltage in accordance with 690.7 at one of the three locations: DC PV disconnect or PV electronic power conversion equipment (inverter usually), or distribution equipment associated with PV system.

Note that there are no specific NEC requirements for the size, color, or wording of this label. Use common sense.

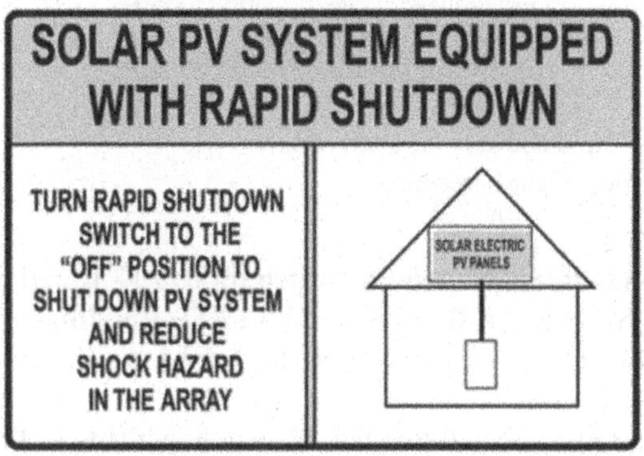

Figure 7.3 Rapid shutdown label.
Source: Courtesy Hellermann Tyton.

690.12(D) Buildings with Rapid Shutdown [in 690.12 Rapid Shutdown of PV Systems on Buildings] (also page 74 of this book)

Rapid Shutdown label

The 2023 NEC Rapid Shutdown (RSD) label has no color requirement, such as with previous NEC versions. Yellow is still acceptable.

The 690.12(D) RSD label location shall be "at each service equipment location to which the PV systems are connected or at an approved readily visible location and shall indicate the location of RSD initiation devices." For specific letter sizes and requirements see page 74.

690.12(D)(1) Buildings with More than One Rapid Shutdown Type [in 690.12 Rapid Shutdown of PV Systems on Buildings / 690.12(D) Buildings with Rapid Shutdown] (also on page 75 of this book, including image)

Sometimes PV systems were installed at different times on a building when there were different NEC rapid shutdown requirements. We do not see this on most PV systems.

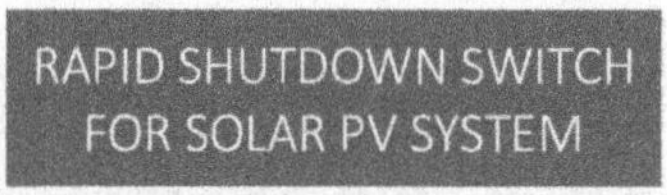

Figure 7.4 Rapid shutdown switch sticker.
Source: Courtesy Eddie Becquerel.

690.12(D)(2) Rapid Shutdown Switch [in 690.12 Rapid Shutdown of PV Systems on Buildings / 690.12(D) Buildings with Rapid Shutdown] (also on page 76 of this book)

This sticker must be within 3 feet of switch, white CAPITALIZED letters, at least 3/8" high on red background.

Figure 7.5 Photovoltaic system disconnecting means label.
Source: Courtesy pvlabels.com.

690.13(B) Marking [in 690.13 Photovoltaic Systems Disconnecting Means] (also on page 81 of this book)

Figure 7.6 Line and load energized in open position sign.
Source: Courtesy Sean White.

The PV system disconnect is usually an ac disconnect and can be the same disconnect as the rapid shutdown initiation device, such as a circuit breaker or a fused disconnect. For a dc interconnected (dc-coupled) system, the PV system disconnect is probably a dc disconnect.

Also, 690.13(B) mentions another labeling requirement, which is often used when it does not need to be, perhaps the most often misused label.

The line and load label is only required to be used in a place where you turn the switch off and can still have voltage on both sides of the switch. This is not required for an interactive inverter output circuit since anti-islanding immediately de-energizes the output of an interactive inverter. If your PV system disconnect is a dc disconnect, then it probably does require this sign. An example of a dc PV system disconnect would be a disconnect at the end of a PV source circuit before it goes into a charge controller. Here you may have voltage from both sides after shutdown. We see this label in different articles that we cover. There are no specific font size or color requirements. Wording can be equivalent.

Figure 7.7 Do not disconnect under load label.
Source: Courtesy pvlabels.com.

690.15(B) Isolating Device [in 690.15 Disconnecting Means for Isolating Photovoltaic Equipment] (also on page 89 of this book)

Here is what it says in the NEC: "Where an isolating device is not rated for interrupting the circuit current, it shall be marked 'Do Not Disconnect Under Load' or 'Not for Current Interrupting.'"

This label is intended for a non-load-break rated isolating device or disconnect, which is used for equipment maintenance. We can see that this is usually marked on PV module connectors when you buy the PV modules. Other places where you need this marking are with finger-safe fuse holders and other non-load-break disconnects.

Figure 7.8 SOLAR PV DC CIRCUIT label.
Source: Courtesy Hellermann Tyton.

690.31(D)(2) Marking and Labeling [in 690.31 Wiring Methods / 690.31(D) Direct-Current Circuits on or in Buildings] (also on page 118 of this book)

The requirement states that we can have the label read either SOLAR PV DC CIRCUIT or PHOTOVOLTAIC POWER SOURCE. We prefer the fewest letters.

This label shall be on exposed raceways, cable trays, other wiring methods, covers or enclosures of pull boxes, junction boxes, or conduit bodies (conduit bodies only if unused available openings).

Color is red on white background with at least 3/8" lettering.

Required for each section separated by enclosures, walls, partitions, ceilings, or floors. Spacing shall not be more than 10 feet.

(No longer a reflective sticker requirement or the word "Warning" required.)

705.10 Identification of Power Sources [Article 705 Interconnected Power Production Sources] (also on pages 169 and 194 of this book)

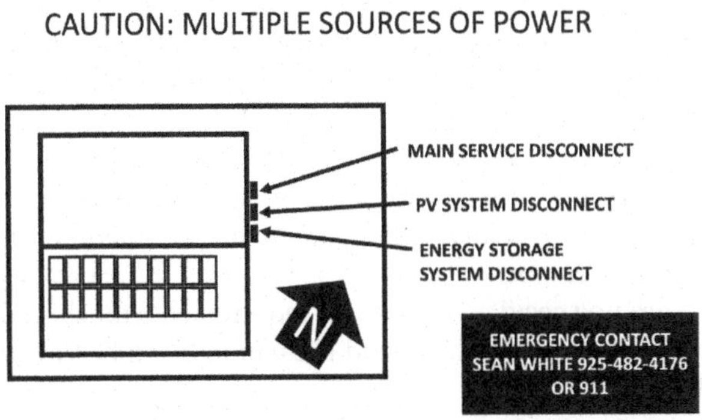

Figure 7.9 Power source placard.

This is a complicated label (placard) and most, if not all, label companies are not doing this by the NEC requirements, which is not unusual.

There are no color requirements; however, you almost always see this one as red with white letters.

The exact words: "CAUTION: MULTIPLE SOURCES OF POWER" are supposed to be on the placard; however, you usually see different words, meaning the same thing.

The new in 2023 NEC requirement is to have an "emergency telephone number of any off-site entities servicing the power source systems."

WARNING: POWER SOURCE OUTPUT DO NOT RELOCATE THIS OVERCURRENT DEVICE

Figure 7.10 120% rule label.
Source: Courtesy Sean White.

705.12(B)(2) and 705.12(B)(4) "120% Rule" label [in 705.12 Load-Side Source Connections / 705.12(B) Busbars] (also on pages 214 and 221 of this book)

It is interesting to note that the suggested wording for the label in the 2023 NEC has one fewer word than in the previous version. The previous version suggests: "WARNING: POWER SOURCE OUTPUT CONNECTION—DO NOT RELOCATE THIS OVERCURRENT DEVICE" and the 2023 NEC is the same, except for removing the word "CONNECTION," which is probably a good thing, since this sticker is often small, so it can fit next to a circuit breaker in a crowded panelboard, so the fewer letters, the bigger they can be.

There are no specific color or size requirements for this label. It just must be adjacent to the circuit breaker and can be the same or equivalent wording.

WARNING: EQUIPMENT FED BY MULTIPLE SOURCES. TOTAL RATING OF ALL OVERCURRENT DEVICES EXCLUDING MAIN SUPPLY OVERCURRENT DEVICE SHALL NOT EXCEED AMPACITY OF BUSBAR

Figure 7.11 Sum rule label.
Source: Courtesy Sean White.

705.12(B)(3) "Sum of the Breakers Rule" label [in 705.12 Load-Side Source Connections / 705.12(B) Busbars] (also on page 217 of this book)

This label is used only if you are using the 705.12(B)(3) Sum of the Breakers method/rule and there are no specific letter sizes or colors called out. Label should be applied to the distribution equipment, with no specifics about where on the distribution equipment.

705.20(7) "Line and Load Energized in Open Position" label [in 705.20 Source Disconnecting Means]

This is repeated in different parts of the NEC. Any time when you open (turn off) a disconnect and there is still voltage on both sides of the disconnect, then you put this label there. We covered this a few pages ago on page 159 in this label section regarding 690.13(B) and Figure 7.6.

705.30(C) Marking [in 705.30 Overcurrent Protection] (also on page 230 of this book)

The simplest thing here to do is quote 705.30(C) in its entirety:

> Equipment containing overcurrent devices supplied from interconnected power sources shall be marked to indicate the presence of all sources.

There are no specifics, so here is a label that is used to fulfill these requirements:

Figure 7.12 Multiple sources label.
Source: Courtesy pvlabels.com.

705.82 Single 120-Volt Supply (also on page 188 of this book)

This label was previously in different articles. If we are using split-phase 120/240V designed equipment for 120V loads that are all in

WARNING:
SINGLE 120-VOLT SUPPLY.
DO NOT CONNECT
MULTIWIRE BRANCH CIRCUITS!

Figure 7.13 Single 120V supply label.
Source: Courtesy Sean White.

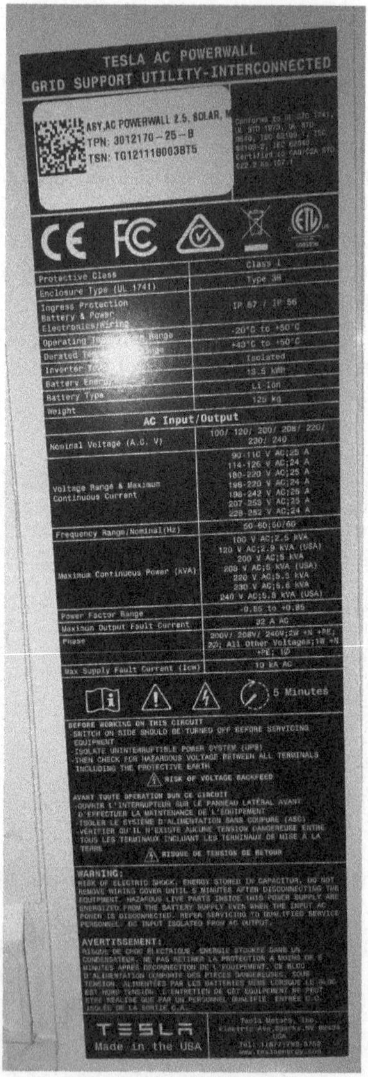

Figure 7.14 Existing label on ESS should be kept visible.
Source: Courtesy Tesla Motors Club discussion forum.

phase with each other, then the neutral can become overloaded if using multiwire branch circuits that were designed for out of phase circuits sharing a neutral. The wording does not have to be exact and can be equivalent. The NEC has an exclamation mark (!), and none of the prominent labeling companies had one, so we made our own!

706.4 System Requirements [Article 706 Energy Storage Systems] (also on page 241 of this book)

These markings are usually going to be **marked on the Energy Storage System**, so you just have to make sure that you do not hide them. The markings include:

(1) Manufacturer's name, trademark, or other descriptive marking by which the organization responsible for supplying the ESS can be identified
(2) Rated frequency
(3) Number of phases, if ac
(4) Rating (kW or kVA)
(5) Available fault current derived by the ESS at the output terminals
(6) Maximum output and input current of the ESS at the output terminals
(7) Maximum output and input voltage of the ESS at the output terminals
(8) Utility-interactive capability, if applicable

706.15(C) Notification and Marking [in 706.15 Disconnecting Means] (also on page 243 of this book)

The ESS disconnecting means label needs to say:
"ENERGY STORAGE SYSTEM DISCONNECT"

(1) Nominal ESS output voltage (this is usually ac voltage coming out of the "system").
(2) Available fault current derived from ESS (this is usually limited by the inverter output and much lower than the available fault current from the utility).

ENERGY STORAGE SYSTEM DISCONNECT
NOMINAL ESS OUTPUT VOLTAGE ------------------------480V
AVAILABLE FAULT CURRENT DERIVED FROM ESS----1000A
DATE CALCULATION WAS PERFORMED-------------1/1/2023

Figure 7.15 ESS disconnecting means label for non-1- and 2-family dwellings (arc-flash label separate here).

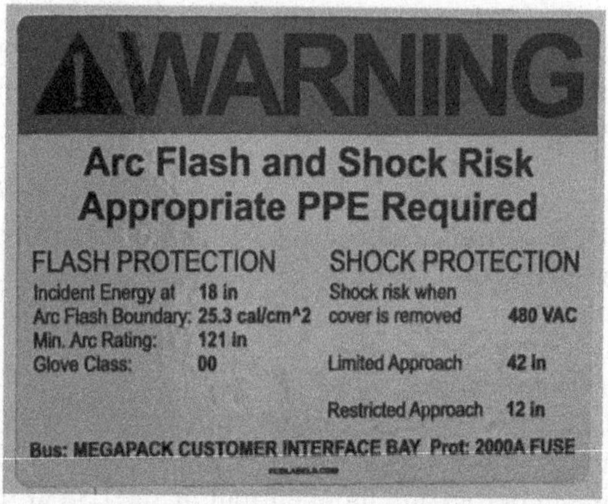

Figure 7.16 ESS Arc-Flash label, Tesla Megapack at Supercharger, Concord, CA.

Source: photo by Sean White.

(3) An arc-flash label applied in accordance with acceptable industry practices (since available fault current is usually limited by electronic power converter, the arc-flash danger is mostly from the utility and this requirement is questionable, unless you have a direct connection to battery cells or a utility-scale ESS).
(4) Date calculation was performed.

There are no 706.15(C) requirements for color or font size of the label. Markings can only be handwritten if the information is subject to change, and with a date calculation was performed requirement—this could be a case where the information is subject to change. *Note that the available fault current, calculations, and arc-flash warnings are a holdover from the days when we were*

Figure 7.17 ESS disconnecting means label for 1- and 2-family dwellings. *Source:* photo by Sean White.

connecting directly to batteries without electronics in between. With the electronics, there is a lot less available fault current and arc-flash hazard. The main hazard is coming from the utility.

Since 2, 3, and 4 above are NOT required for 1- and 2-family dwellings, then here is what your label will look like for your duplex or home.

One more thing about 706.15(C) ESS disconnect labeling: the line and load energized sticker, which you saw on page 159, Figure 7.6, is only required if there can be voltage on both sides of the disconnect after the switch is off. So for an interactive inverter only circuit, this sticker is not required, unless your ESS can operate in backup mode. Then you probably can have voltage on both sides after opening the disconnect.

A Word about 110.21(B) Field Applied Hazard Markings

Many of our labeling requirements reference 110.21(B), which pretty much says that we need to use labels "of sufficient durability to withstand the environment involved and warn of the hazards using effective words, colors, symbols, or any combination thereof." There are no specific requirements here; however, there is an Informational Note that directs us to ANSI Z535.2-2011, which can give us ideas about colors and symbols. Recall that Informational Notes are not requirements; however, the AHJ may require whatever they want, since they have Zeus-like powers. (If your inspector has Dionysus-like powers, offer them a bottle of wine in order to pass.)

Additionally, 110.21(B) tells us to permanently affix the labels and not to use handwriting, unless the information is subject to change. *Does this mean that if the information is not subject to change, we can put the pen in our teeth and write as long as we don't use our hands?*

706.15(E)(3) Remote Activation [in 706.15 Disconnecting Means / Disconnecting Means for Batteries] (also on page 245 of this book)

This requirement is only for cases where there are no electronics (inverters or dc-to-dc converters) associated with the batteries and where a remote switch for a disconnecting means is used. **This is not common** and perhaps there is no such thing as a UL 9540 listed system of this type. In this case, on the disconnecting means itself, you put a label indicating where the remote switch is located. There are no special requirements for the wording, font, or anything else about this label. We recommend plaid labeling (joke).

706.15(E)(4) Notification [in 706.15 Disconnecting Means / Disconnecting Means for Batteries] (also on page 245 of this book)

As with 706.15(E)(3), this is only for where you have high available fault currents, like batteries without electronics, not your typical ESS. This label has the **same requirements as the ESS disconnect label we recently covered on page 166 and in Figures 7.15 and 7.16, except we do not mark it as an "ENERGY STORAGE SYSTEM DISCONNECT,"** and there are not exceptions for 1- and 2-family dwellings.

The International Residential Code (IRC) requires a UL 9540 listed ESS, and you probably are not going to find one that does not have electronics protecting the batteries, unless it was listed after the publishing of this book.

706.21(A) Facilities with Utility Services and ESS [in 706.21 Directory (Identification of Power Sources)] (also on page 248 of this book)

706.21(A) sends you to **705.10 Identification of Power Sources**, which is on page 160 in this marking and labeling section of this chapter. Essentially, you need to identify the different power sources.

706.21(A) Facilities with Stand-Alone Systems [in 710.10 Identification of Power Sources] (also on page 248 of this book)

We will show an image of this label coming up on page 252 when we cover Article 710 Stand-Alone Systems.

706.41 Electrolyte Classification [in 706 Part V Flow Batteries] (also on page 250 of this book)

Electrolytes shall be identified by name and chemical composition.

710.10 Identification of Power Sources [in 710 Stand-Alone Systems] (also on page 252 of this book)

Label shall be at power source disconnecting means or an "approved readily visible location" and shall include the location(s) of power source disconnecting means for building.

If there are multiple sources, such as a PV system and an ESS, then markings will comply with the 705.10 requirements covered on pages 160 and 194 and Figure 7.9. In most cases you are

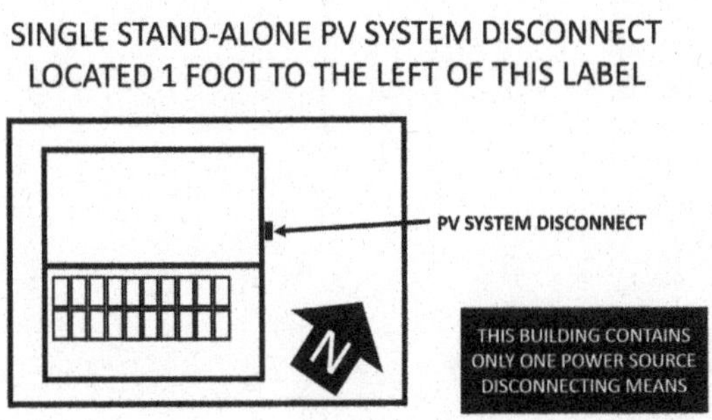

Figure 7.18 Stand-alone single power source label.

going to have more than one source; however, if you have a single multimode inverter with PV and battery inputs and a single dc disconnect, then perhaps you can try and consider it one source. A direct water pumping system on a building would qualify as a single power source, but not something you will see every day or ever. Perhaps 710.10 can be made better next time around. We recommend using the 705.10 label and ignoring Figure 7.18 unless you have a direct PV pump house with no other power source.

WARNING:
SINGLE 120-VOLT SUPPLY.
DO NOT CONNECT
MULTIWIRE BRANCH CIRCUITS!

Figure 7.19 No multiwire branch circuits label for stand-alone 120V inverter panelboards.

710.15(C) Single 120-Volt Supply [in 710.15 General] (also on page 255 of this book)

This is only when you supply 120/240V distribution equipment with a 120V stand-alone inverter.

8 Article 691 Large-Scale Photovoltaic (PV) Electric Supply Stations

Article 691 was new in the 2017 NEC and there is not a whole lot to it (about one page in the NEC). There are certain things that can be done with these larger systems, under the supervision of an engineer, which cannot be done with smaller systems.

In the past, our excuse for changing the rules on these large "utility-scale" solar farms was that we decided to call them utilities, and utilities are not subject to the requirements of the NEC. Utilities typically use engineering standards and the **National Electrical Safety Code (NESC)** for designing their systems. The question has always been: Is it really a utility? Now, we can use the NEC and no longer have to look for a "utility" loophole when installing large-scale PV. Also, many jurisdictions did not buy that a large-scale PV system was like utility-owned properties and would enforce Article 690. Many local AHJs do not have the training and experience to inspect a large powerplant and are more qualified to inspect distributed generation PV systems, such as PV systems on or near buildings.

Not being able to treat a large solar facility like a large coal, nuclear, or natural gas facility, when it comes to the local AHJ and the NEC, was discriminatory against renewable energy. Finally, the NEC woke up!

Outline of Article 691 Large-Scale PV

691.1 Scope (Large-Scale PV not under utility control)
  691.1 Informational Note 1: sole purpose to supply utility
  691.1 Informational Note 2: refers to 90.2(B)(5) and NESC

DOI: 10.4324/9781003189862-9

691 Large-Scale Photovoltaic (PV) Electric Supply Stations

A 5+MWac PV system can be installed outside of the scope of 691, for instance, on a building. Just because a PV system is larger than 5MW does not automatically make it compliant with 691. It must meet all of 691.4's special requirements, with being over 5MWac just one of the requirements. (See page 175.)

691.1 Scope

Article 691 only applies to **PV systems not under exclusive utility control**.

Note: In the 2020 NEC 691.1 was where we were told that the PV system had to be over 5MWac and this part was moved to 691.4(7) (page 178).

Section 691.1 **Scope** mostly consists of Informational Notes, including an image, which is a type of Informational Note here. Recall that Informational Notes are not enforceable parts of the Code and are just good ideas.

691.1 Informational Note 1 tells us that these 691 systems are unique, are **only to supply a regulated utility**, and to look to 691.4 Special Requirements for Large-Scale PV Electric Supply Stations (page 175).

691.1 Informational Note 2 tells us that utility-owned properties are not the domain of the NEC (even 691) and to see the **National Electrical Safety Code (NESC)**.

There is also a **typo in 691.1 Informational Note 2**. Like the 2020 NEC, it refers us to **Section 90.2(B)(5)**. There is no longer a **90.2(B)(5)** in the 2023 NEC, since it was moved to **90.2(D)(5)**. **In the 2023 NEC it should have said 90.2(D)(5)** rather than 90.2(B)(5).

Get your bearings in Article 90 here:

- Article **90 Introduction**
- Section 90.2 **Use and Application**
- 2023 NEC **90.2(D) Installations Not Covered** tells us the NEC does not cover certain installations, like cars, mobile homes, boats, trains, etc.

90.2(D)(5), like **90.2(B)(5) in the 2020 NEC**, tells us that **installations under exclusive control of the utility do not follow the NEC**, such as service drops, transmission, distribution, metering, **generation**, etc.

NESC

There is a reference in **691.1 Informational Note No. 2** to the **National Electrical Safety Code (NESC)**, which is also known as **ANSI Standard C2** and is published by IEEE.

- ANSI is the American National Standards Institute. ANSI accredits standards that are developed by other standards organizations, government agencies, consumer groups, companies, and others.
- IEEE is the Institute of Electrical and Electronics Engineers. World's largest association of technical professionals with more than 423,000 members in over 160 countries.

Discussion: Most utility-owned PV supply stations look just like every other PV power plant that uses Article 691, although they are not using the NEC. This makes equality for utility-owned and non-utility-owned large-scale PV. Perhaps this equality for non-utility-owned PV should also be the 28th amendment to the US constitution.

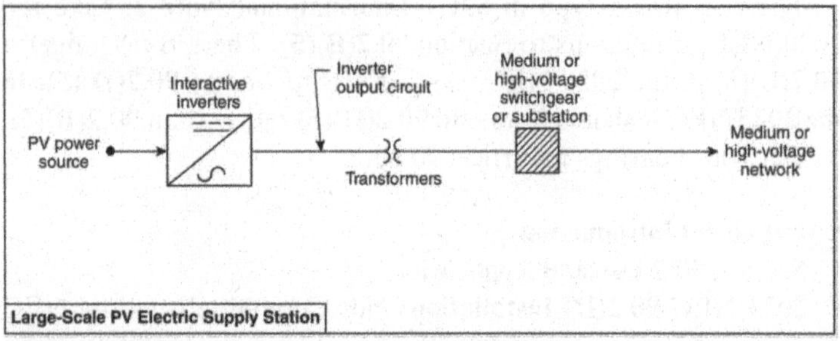

Figure 8.1 Informational Note Figure 691.1 Identification of large-scale PV electric supply station components.

Source: Courtesy NFPA.

Discussion: The image in 691.1, which is the only image in Article 691, is showing a single-line diagram (SLD) of a dc PV power source going to an inverter and then to a medium or high voltage transformer with the switchgear (overcurrent protection and disconnecting means) on the medium/high voltage side of the transformer. Medium voltage is typically many thousands of volts; however, there are different definitions of how low, medium, and high voltage are defined. In the 2020 NEC, this "informational" image was slightly different and where it now says, "medium or high-voltage substation or switchgear," it used to say, "substation optional." Additionally, where it used to say, "electric production and distribution network," it now says "medium or high-voltage network." These changes are for clarity and make no substantial difference.

High-Voltage and Medium-Voltage Benefits

Oftentimes the utility-scale solar plant will have medium-voltage connections inside the plant and then all the medium-voltage cables will combine at a substation where the voltage will be boosted to high voltage. An example would be medium voltage of 34.5kV and high voltage of 115kV. At 115kV a MW would be 1000kW/115kV/1.732 = 5A. As you can see, with higher voltage, the current goes way down!

691.4 Special Requirements

Section 691.4 stayed mostly the same as in the 2020 NEC; however, 691.4(6) and 691.4(7) were added. 691.4(6) is a new requirement for a "central command center," which sounds pretty cool. 691.4(7) was moved from 691.1. 691.4 is the longest Section in a short Article.

691.4(1) Qualified Persons

Electrical circuits and equipment shall be maintained and operated only by qualified persons.

Discussion: Qualified person as defined in NEC Article 100:

> One who has skills and knowledge related to construction and operation of electrical equipment and installations and has received safety training to recognize and avoid the hazards involved.

This means that there is no hard definition of qualified persons; however, common sense should prevail. Some people consider NABCEP certification as a way of showing that someone is qualified, and others believe that an electrician is qualified personnel. Perhaps only people reading this book are truly qualified. It would also depend on the job someone is doing. The average person who is qualified to install solar racking systems at a large solar plant is likely not qualified to work on medium-voltage or high-voltage equipment.

It is interesting that 691.4(1) Informational Note directs us to **NFPA 70E Standard for Electrical Safety in the Workplace** for the qualified personnel definition and that it is the same exact definition that we find in NFPA 70, the NEC. Perhaps NFPA is doing a good job at trying to sell more books. We should learn from them!

691.4(2) Restricted Access

Access to PV electric supply stations shall be restricted by fencing or other means in accordance with 110.31. Additionally, field-applied hazard markings shall be applied in accordance with 110.21(B).

110.31 Enclosure for Electrical Installations is in Article 110 Requirements for Electrical Installations **Part III Over 1000V, Nominal** and this section of the Code goes into the details of how to keep unqualified people out of the electrical installation.

Some of the other means besides fencing for keeping the PV area restricted to qualified people according to 110.31 are:

- vaults
- rooms
- closets
- walls
- roofs
- floors
- doors
- locks

691.4(2) also directs us to **110.31 Enclosure for Electrical Installations** and here we find Table 8.1.

Table 8.1 Table 110.31 Minimum distance from fence to live parts

Nominal voltage	*Minimum distance to live parts*	
	m	*ft*
1001–13,799	3.05	10
13,800–230,000	4.57	15
Over 230,000	5.49	18

Source: Courtesy NFPA.

691.4(3) Medium- or High-Voltage Connection

The connection between the PV electric supply station and the utility shall be made through a **medium-voltage or high-voltage method**, such as the following:

- switchgear
- substation
- switchyard
- similar method

There are different definitions of medium- and high-voltage, but a general consensus is that medium voltage is over 1000V, and here we are typically talking about connections that are **between 4000 and 500,000V**.

That means no 480V 5MW projects here. That would be a lot of current.

5MW / 480V / 1.732 for 3-phase = over 6000A!

691.4(4) Loads Only for PV Equipment

The only loads allowed in a PV electric supply station are those to power the auxiliary equipment used in the process of generating power.

Examples of loads would include monitoring equipment, weather stations, lights, controls, and PV maintenance equipment. (I guess that means no 5MW cell phone charging. Sorry if you bought that new HVDC 5MW USB-C charger, sucker.)

691.4(5) Not Installed on Buildings

Article 691 will not apply to PV systems installed on buildings. This does not mean that PV systems larger than 5MW cannot be on a building; it means that the special rules in Article 691 will not apply to any PV system installed on a building. Article 690 applies in all cases where Article 691 does not apply for PV and the NEC.

Question: If you had 5MW over an underground Y2K bunker, would it be "on a building"?—I don't think so, Tim.

691.4(6) Central Command Center

"The station shall be monitored from a central command center."

Discussion: This is new in the 2023 NEC. There is no NEC definition of central command center. We are assuming that it has nothing to do with NATO and that the reason this is here is that Bill is a Spaceballs movie fan.

691.4(7) >5MW

Invoking Article 691 can only be done with systems with an inverter generating capacity of at least 5000kW. There is also an Informational Note clarifying that individual sites operated as a group with a total generating capacity of 5000kW fit this definition.

691.4(7) is nothing new and was formerly in 691.1 Scope in the 2020 NEC.

Article 100 Definition: Generating Capacity, Inverter

> The sum of the parallel-connected inverter maximum continuous output power at 40°C in watts, kilowatts, volt-amperes, or kilovolt-amperes.

Essentially this is the ac output of the inverter. New to the 2023 NEC is the definition of Inverter Generating Capacity expanded to include apparent power and not just real power. Apparent power is often expressed in kVA and is different from real power when your power factor is less than 1.

Oftentimes the dc PV portion of a PV system is sized between 1.2 and 1.5 times greater than the inverter generating capacity for large-scale PV projects. We have seen as high as a 2:1 ratio!

The NEC likes to use kW rather than MW for big power, but we like to use MW, since it sounds more big time!

691.5 Equipment

Equipment shall be approved by one of the following methods:

691.5(1) Listing and Labeling

A listed PV module tested to UL 61730 and labeled would be an example.

691.5(2) Field Labeling and Identified for Application

Typically, field labeling would be having a Nationally Recognized Testing Lab (NRTL), such as UL, TUV, CSA, or Intertek send someone out to approve of the product and give it the lab's label while it is in the field. These field labels may represent a subset of the tests that are conducted on factory labeled equipment since not all tests are feasible in the field.

691.5(3) Engineering Review

When listing and labeling or field labeling are not available, an engineering review can take place to validate that the equipment is evaluated and tested to an industry standard or practice. Keep in mind **this method is only available if the other two methods are *not* available**. It is up to the AHJ whether to approve of this process.

691.6 Engineered Design

Documentation stamped by a licensed professional engineer shall be made available at the request of the AHJ.

The engineer shall be independent and retained by the system owner.

Additional stamped engineering reports shall be made available upon request of the AHJ, documenting:

- Compliance with Article 690
- Alternative methods used not in compliance with Article 690
- Alternative methods used not in compliance with the NEC
- Compliance with industry practice

Discussion: What this means is that under engineering supervision, we are allowed to stray from Article 690 when we meet the requirements of Article 691. Section 691.6 is a documented version of what is commonly performed during the plan check phase of a construction project. Several examples of typical areas where a large-scale PV system is likely to take exception to Article 690, or the rest of the NEC, are listed in 691.8 through 691.11.

691.7 Conformance of Construction to Engineered Design

Documentation that **construction of the project followed** the electrical engineered **design** shall be made available to the AHJ upon request.

Additional licensed professional electrical engineer stamped reports detailing that **construction conforms with the NEC standards and industry practice** shall be provided to the AHJ upon request.

The engineer shall be independent and retained by the system owner.

Discussion: Section 691.7 is a documented version of what is commonly performed during the field inspection phase of a construction project.

691.8 Direct-Current Operating Voltage

Included in the documentation required by **691.6 Engineered Design** shall be voltage calculations.

691.9 Disconnecting Means for Isolating Photovoltaic Equipment

Isolating devices (non-load-break rated disconnecting means) shall be **permitted to be remote from the equipment**. (**690.15 Disconnecting Means for Isolating Photovoltaic Equipment / 690.15(C) Equipment Disconnecting Means** requires isolating devices or remote operating devices to be 10 ft from equipment and 691 can let us ignore this requirement.)

Put in engineered design if disconnecting means remote from equipment:

- Written safety procedures ensure only qualified persons service equipment.
- Maintenance conditions ensure only qualified persons service equipment.
- Supervision ensures only qualified persons service equipment,

Informational Note: **Lockout-Tagout** procedures are in **NFPA 70E Standard for Electrical Safety in the Workplace**. For maintenance information, see **NFPA 70B Recommended Practice for Electrical Equipment Maintenance**.

Adherence to **690.12 Rapid Shutdown is not required** for buildings whose sole purpose is to protect PV supply station equipment.

Written standard operating procedures detailing shutdown procedures in case of emergency shall be available on site.

691.10 Fire Mitigation

691.10 was changed from Arc-Fault Mitigation to Fire Mitigation, which was the whole point of arc-fault mitigation, to prevent fires. No other wording was changed in 690.10, except for the addition of a new Informational Note. This **Informational Note tells us that fire mitigation plans are typically reviewed by the local fire agency and include things such as local access roads.**

If PV system does not comply with 690.11 Dc Arc-Fault Protection, then included in the documentation in **691.6 Engineered Design** shall be **fire mitigation plans to address dc arc-faults**.

Example of Fire Mitigation Plans

- Include specific details for firefighting within the PV plant.
- Include access in and around the PV plant for fire department equipment.
- May provide on-site suppression capabilities including extinguishers at each inverter pad and tanker trucks with fire hoses for early response.
- May include personnel on site during operating hours that can see fires as they get started.
- "Hot Work" practices followed for cutting and welding within the facility.
- ("Hot Work" is the main cause of fires in a large-scale PV facility.)

691.11 Fence Grounding

Fence grounding requirements and details shall be included in the documentation in **691.6 Engineered Design**.

691.11 Informational Note

Directs us to:

250.194 Grounding and Bonding of Fences and Other Metal Structures

250.194 relates bonding and grounding of fences to substations, and the Informational Note does not mean we are required to follow 250.194; however, it is a good reference. We are looking out here for things such as step voltages, which means that in certain instances you could take a step and the voltage could be different one foot to the next if a power line hit the ground. Ouch! (This means that electricians with shorter legs are inherently safer.)

250.194(A) Metal Fences goes on to say that we need to bond fences that are within 16 ft of exposed electrical conductors or equipment.

- **250.194(A)(1)** Bonding jumpers are supposed to be every 160 ft and at fence corners.
- **250.194(A)(2)** When bare overhead connectors cross over a fence, bonding jumpers shall be on each side of the crossing.
- **250.194(A)(3)** Gates shall be bonded to gate post which shall be connected to electrode system (people have been shocked by opening a gate if different sides of the gate are not bonded to each other).
- **250.194(A)(4)** Gate or opening in fence shall be bonded by buried bonding jumper.
- **250.194(A)(5)** Grounding grid or electrode systems shall cover the swing of gates.
- **250.194(A)(6)** Barbed wire strands above the fence bonded to the electrode system.

Question: Why ground the barbed wire? To protect the criminals? How about a medium-voltage electric fence to keep out the riffraff?

Article 691 Overview

When installing a large-scale PV system in the past, PV companies were often forced to pretend that they were a utility. Using Article

690 for compliance with a 100MW power plant was never the intended use of Article 690. These large systems are unlike smaller systems, in that they are not accessible to the public, as are PV systems on buildings or in backyards. With the ability to do things under engineering supervision that stray from the requirements of Article 690, we are no longer put into a position where we have to pretend that we are a utility, and that the system is behind a utility fence.

Utilities may still build and operate PV systems that are not required to be compliant with the NEC; however, it is now Code-compliant to build a large PV system that does not comply 100% with Article 690. This flexibility can actually improve operation, maintenance, and safety for these large power plants.

9 Article 705 Interconnected Electric Power Production Sources

As both Articles 690 and 705 developed over time, the requirements in 690.64 were incorporated into Article 705 in the 2011 NEC. With the advent of much more distributed generation coming onto the grid from solar PV, wind, and other sources, Article 705 has grown in size and importance.

In this chapter, our main focus will be on the most important solar PV and energy storage related material in Article 705, which is connecting utility interactive inverters to the grid. However, microgrids are becoming more important, so another important focus will be ac and dc microgrids.

In the 2020 and 2017 versions of the NEC, there was Article **712 DC Microgrids**, which was **removed in the 2023 NEC**, and **we now look to Article 705 for dc microgrids**. See microgrid definition on page 267.

Many of the requirements of Article 705 are satisfied by the listing of the interactive inverters.

Reviewing an outline of an article before studying the article helps us get properly situated and our minds organized to study the article.

Outline of Article 705 Interconnected Electric Power Production Sources

705 Part I. General
- 705.1 Scope
- 705.1 Informational Notes and Figures
  - DC Interconnected Image
  - AC Interconnected Image

DOI: 10.4324/9781003189862-10

- 705.5 Parallel Operation
  - 705.5(A) Output Compatibility
  - 705.5(B) Synchronous Generators
- 705.6 Equipment Approval
- 705.8 System Installation
- 705.10 Identification of Power Sources
  - 705.10(1) Disconnecting Means Location
  - 705.10(2) Emergency Phone Number
  - 705.10(3) "CAUTION: MULTIPLE SOURCES OF POWER" Label
- 705.11 Source Connections to a Service
  - 705.11(A) Service Connections
    - 705.11(A)(1) New Service
    - 705.11(A)(2) Supply Side of Service Disconnecting Means
    - 705.11(A)(3) Additional Set of Service Entrance Conductors
  - 705.11(B) Conductors
    - 705.11(B)(1) Output Rating (Ampacity ≥ Sum of Maximum Circuit Current)
    - 705.11(B)(2) ≥ 6 AWG Copper or 4 AWG Aluminum
    - 705.11(B)(3) Ampacity of Other Service Conductors ≥ (B)(1) and (B)(2)
  - 705.11(C) Connections
    - 705.11(C)(1) Splices or Taps
    - 705.11(C)(2) Existing Equipment
    - 705.11(C)(3) Utility-Controlled Equipment
  - 705.11(D) Service Disconnecting Means
  - 705.11(E) Bonding and Grounding
  - 705.11(E) Overcurrent Protection
- 705.12 Load-Side Source Connections (half of our 705 focus is here)
  - 705.12(A) Feeders and Taps
    - 705.12(A)(1) Feeder Ampacity ≥ 125% Inverter Current
    - 705.12(A)(2) Load side of Inverter Connection to Feeder
      - 705.12(A)(2)(a) Feeder Ampacity ≥ Feeder OCPD + 125% Inverter Current
      - 705.12(A)(2)(b) OCPD ≤ Feeder Ampacity on Load Side of Connection Point

705.12(A)(3) Tap Conductors use 1/3 of feeder breaker + 125% Inverter Current
705.12(B) Busbars
705.12(B)(1) 100% Rule
705.12(B)(2) 120% Rule
705.12(B)(3) Sum Rule
705.12(B)(4) Center-Fed 120% Rule
705.12(B)(5) Feed-Through Conductors
705.12(B)(6) Switchgear, Switchboards and Busbars with Engineering Supervision
705.13 Energy Management Systems (formerly Power Control Systems)
705.20 Source Disconnecting Means
705.20(1) Must be one of the Following
705.20(1)(a) Manually Operated Switch or Breaker
705.20(1)(b) Load-Break-Rated Pull-Out Switch
705.20(1)(c) Power Operated Remote-Controlled Switch
705.20(1)(d) Device Listed for Application
705.20(2) Simultaneously Disconnect Ungrounded Conductors
705.20(3) Readily Accessible
705.20(4) Externally Operable Without Exposed to Live Parts
705.20(5) Indicate On or Off
705.20(6) Ratings Sufficient Maximum Circuit Current, Fault Current and Voltage
705.20(7) Line and Load May be Energized in Open Position Sign
705.25 Wiring Methods
705.25(A) General
705.25(B) Flexible Cords and Cables
705.25(C) Multiconductor Cable Assemblies
705.28 Circuit Sizing and Current
705.28(A) Power Source Output Maximum Current
705.28(A)(1) Sum of Power Production Equipment Currents
705.28(A)(2) Current Setpoint of Energy Management System (EMS)
705.28(A)(3) Sum of 705.28(A)(1) + 705.28(A)(2)
705.28(B) Conductor Ampacity
705.28(B)(1) 125% of Maximum Currents
705.28(B)(1) Exception 1: Listed for 100%

705 Part III. Interconnected Systems Operating in Island Mode (New Part/Organization)
705.80 Power Source Capacity
705.81 Voltage and Frequency Control
705.82 Single 120-Volt Supply with Label

705.1 Scope

Article 705 covers the installation of **multiple power sources connecting in parallel**, such as from a renewable energy source and the utility. One of the power sources must be a primary power source.

Usually, it is the grid that is the primary power source, but there can be exceptions, such as **when Chuck Norris connects to the grid, the grid becomes secondary**. This last statement was a joke, but we would like to think that, as we are producing and storing our own energy more and more, the grid sometimes is the smaller part of the power and energy that we are using.

705.1 Informational Notes, Including Informational Note Figures

A very noticeable change in Article 705 is the movement of the images for ac-coupled and dc-coupled systems from the beginning of Article 690 to the beginning of Article 705. These images are still for informational purposes only, and have changed somewhat, besides being moved. Even the terms ac-coupled and dc-coupled have changed to ac-interconnected and dc-interconnected.

705.1 Informational Note 1: A Primary Power Source May Be a Utility or an On-site Electric Power Source

Discussion: The utility part is obvious; however, if there is no utility, then something must be primary. We may look at this "primary" as the drummer of a rock band making the beat (frequency). If we had designed an off-grid system with multiple inverters in the past, the primary power source may have been called a "master" and the secondary something even more offensive that we cannot say in this book, or else we would get kicked off Twitter (at least in the good old days). We now have "less offensive to everyone" terms called "primary power source" and "secondary primary source." Once you are in island-mode with your microgrid, you need to assign something to be primary. A simple example of this is with a small microgrid with a single battery inverter that is the primary power source working with interactive PV inverters, which follow the primary inverter's frequency. If you have multiple battery inverters in your microgrid, then you have to assign one to be primary. You cannot have two frequencies in a microgrid, which would be like having two people steering a sailboat in different directions, or to use a musical analogy, good jazz.

705.1 Informational Note Figure(s)

These are the images that were modified and moved from Article 690 in the 2023 NEC. Let us first look at the images (Figures 9.1 and 9.2 on the following two pages) and then talk about them. The two images are treated as one in the 2023 NEC.

Difference in 2023 and 2020 NEC images (recall that these images are for informational purposes only and there are many variations so it is unlikely that your system will look exactly like this):

- Dc coupled multimode system changed to **DC interconnected example**
- Ac coupled multimode system changed to **AC interconnected example**
- Microgrid Interconnect Device (MID) is not in 2020 NEC image (See MID definition on page 267)
  - In dc interconnected example **MID is internal to inverter**
  - In ac interconnected example **MID is external to inverter**
- PV power source changed to **power source**
- PV system disconnect changed to **source disconnect**
- Interactive disconnect changed to **source disconnect**
- Electric power production and distribution network changed to **primary source**
- Dedicated loads changed to **ac loads**

Informational Note Figure 705.1 Discussion: **Changing coupled to interconnected**: This will be a difficult change in terminology, since we have been using the term for so long. Coupled will probably be

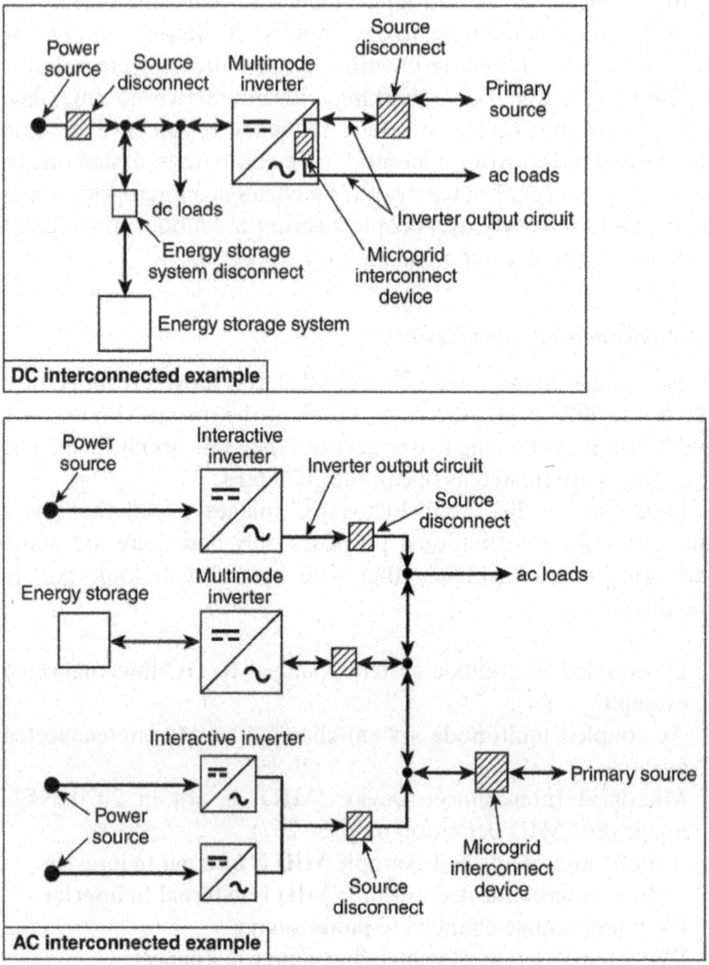

Figure 9.1 2023 NEC Informational Note Figure 705.1 identification of power source components in common configurations.

Source: Courtesy NFPA.

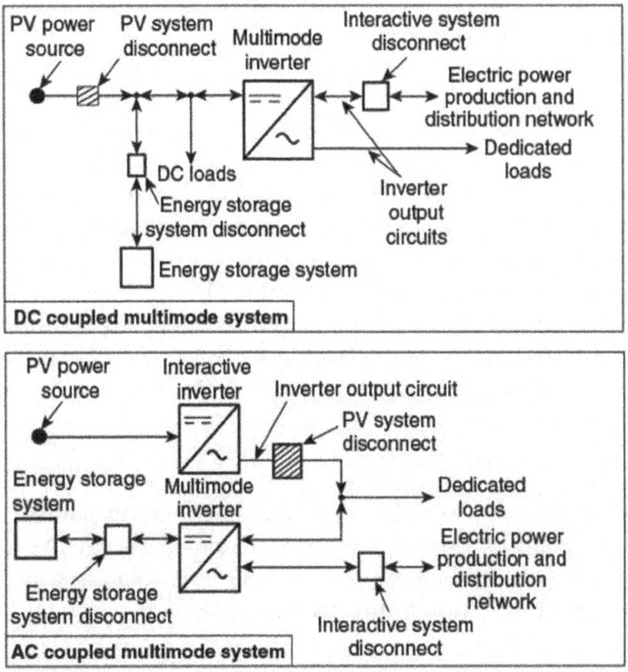

Figure 9.2 2020 NEC Figure 690.1(A) (part of image).
Source: Courtesy NFPA.

used in the industry, just like solar panel, which is also no longer defined or used in the NEC.

MIDs in the image: MIDs were always in your typical multimode inverter, even before the term MID was in the NEC. This was how your multimode inverter could disconnect from the grid and switch from interactive-mode to island-mode when the grid goes down before we were using external MIDs. Recall that **these images are examples only and an ac interconnected system may have its MID internal to an inverter, and a dc interconnected system may have an external MID.**

It appears that the images being moved to Article 705, and the modifications to the images, reflect energy storage being more popular. The Code is being more inclusive of different power sources being connected together, and not being as PV-centric, although we know that PV is most important, at least until we can control small fusion reactions with banana peels.

705.5 Parallel Operation

705.5 Parallel Operation **was 705.14 Output Characteristics in the 2020 NEC** and was moved around and reworded, with a list in it for the 2023 NEC, just to keep us on our toes. There is no difference in how we implement this material.

705.5 should be covered by the UL 1741 listing of the inverter, which is also mentioned in an Informational Note. Interactive inverters installed in the US must be listed to UL 1741 and satisfy the requirements of 705.14.

705.5(A) Output Compatibility

Power production sources (inverters) operating in parallel shall have compatible voltage, wave shape, and frequency. As a side note, a power production source could be a rotating generator; however, very few non-inverter technologies have a UL 1741 listing.

Discussion: If this were not the case, a UL 1741 inverter in interactive mode would not turn on. There was an Informational Note in the 2020 NEC that said **being compatible with the wave shape does not mean matching the wave shape**. Sometimes we will want to modify the wave shape to condition the power and make it better. For instance, many utility-scale inverters are required to make reactive power, which means making a wave shape different than what is on the grid. Many smaller-scale PV systems, such as the one on your house, will make perfect power factor to a grid that has an imperfect wave form. Perhaps this Informational Note did not need to be in the 2023 NEC, since after reading this book, the industry found it obvious that the wave shape does not have to match. Another example of a wave shape not matching, would be an inverter injecting reactive power into the grid, to improve the power quality and transmission capacity. Reactive power can support the voltage when electricity is transmitted over long distances, and for utility-scale PV systems, there are requirements for providing reactive power.

Wave Shaping for a Better Grid with Inverters That Work at Night

There are some inverters that are being modified to condition the power quality on the grid even at night when there is no solar energy available. Since we are putting the solar inverters on the grid, it does not take a lot to modify the inverters to condition the grid. All we need is the policy in place to make it feasible and in the UK an organization called Lightsource BP is doing this. Get ready for more in the future, especially when there is energy storage involved.

Reference: www.greentechmedia.com/articles/read/lightsource-bp-makes-solar-pay-at-night

705.5(B) Synchronous Generators

Synchronous generators shall be installed with the required synchronizing equipment.

705.6 Equipment Approval

Equipment shall be **approved for intended use**. We need to use equipment that is approved for interconnection to the grid or to operate in parallel with other power production sources.

Equipment that should be listed interactive or field labeled and evaluated interactive includes:

- interactive inverters
- engine generators with special controls
- energy storage equipment
- wind turbines

Recall that the listing standard for an inverter is UL 1741.

705.6 Informational Note 1

Reference to UL 1741, Standard for Inverters, Converters, Controllers, and Interconnection System Equipment. Stand-alone, island, interactive, and multimode:

- stand-alone sources = island mode
- interactive sources = interactive mode
- multimode sources = island mode or interactive mode

705.6 Informational Note 2

Interactive functions are also common in:

- microgrid interconnect devices (MID)
- power control systems (PCS)
- interactive inverters
- synchronous engine generators
- ac energy storage systems
- ac wind turbines

705.8 System Installation

Installations only to be performed by qualified person (see pages 16 and 175 of this book for discussion on qualified person). Or see Article 100 Definition of Qualified Person.

705.10 Identification of Power Sources

Permanent **plaques, labels, or directories** shall be installed at each service equipment location or at an approved readily visible location in accordance with 705.10(1) through (3) below.

705.10(1) Disconnecting Means Location

Denote each power source disconnecting means location.

Discussion: This means that if you have a separate disconnecting means in different locations for three PV systems, two energy storage systems, one Ford F-150 bidirectional EV Supply Equipment (charger/discharger), and one service disconnect, then you will need seven signs in seven locations. This is nothing new and is a theme.

705.10(1) Exception

If multiple disconnecting means are in one location, then one sign per location is sufficient.

705.10(2) Emergency Phone Number

Have emergency phone numbers of off-site entities servicing power systems.

Discussion: This is a new requirement to provide contact information when those responsible for the operation of the equipment are off-site. This would also include companies that have a service agreement to maintain the system who also must respond if the equipment malfunctions. Since many systems require specialized technicians, it was decided that this language would help emergency personnel locate qualified technicians to help during an emergency where the equipment is involved. This is new!

705.10(2) Informational Note

We are referred to 2021 NFPA 1 Fire Code (which is not adopted in most states, since most states adopt the International Fire Code) and here is all it says:

> **11.12.2.1.5 Installer Information**
>
> A label shall be installed adjacent to the main disconnect indicating the name and emergency **telephone number** of the company currently servicing the PV system.

This sounds like a good idea. Perhaps every fire department should have a phone number of a solar and storage "qualified person" nearby, so they can jump out of bed and share knowledge, which can save lives. Do not go putting your ex's phone number on that label, unless they are "qualified" (sing 867-5309).

705.10(3) "CAUTION: MULTIPLE SOURCES OF POWER" Label

"CAUTION: MULTIPLE SOURCES OF POWER" shall be marked on the label in accordance with 110.21(B). 110.21(B) is a common reference for labeling, which essentially tells us that the label needs to last a long time and not be handwritten.

If there is an emergency, the firefighters would like to know the location of every power source disconnecting means, so they can know when they turn everything off. This plaque, label, or directory must be displayed at every power source disconnect location, and must show all the other power source disconnect locations. Firefighters do not want to think they have turned off the building only to find out that they missed a power source disconnect hidden in the backyard. The information in 705.10 has been an important theme in the NEC throughout the years.

The NEC has no definitions of plaque, "label," or directory and added label to 705.10 in the 2023 NEC.

Interconnections 705.11 and 705.12

Part of the magic of PV systems is connecting inverters to the grid and sending power backward. For an electrician who is used to power always coming from one direction, this can be new and exciting. We can have fewer currents coming from the utility and busbars fed from both ends. In a way, this is how the grid looks to a utility with multiple power plants. Hopefully they use the 120% rule when connecting a nuclear powerplant to the grid.

Time to dive in!

705.11 Source Connections to a Service (Formerly Supply-Side Source Connections)

In the 2023 NEC, we see the former 705.11 "Supply-Side Source Connections" changed to "**Source Connections to a Service**," so we no longer have "Supply-Side Connections" and we never had "Line-Side Taps" in the NEC, but we can still say these terms offline, or perhaps we can just take the hyphen out between supply and side. Besides renaming, mostly rearranging and making lists are the 705.11 2023 NEC changes.

A source connection to a service (formerly supply-side connection) is the connecting of a parallel power source, such as a solar interactive inverter, on the **supply side of all overcurrent protection** on a service. **If the inverter is on the load side of any over-current protection**

protecting loads, then it is not a supply-side connection. Typically, supply-side connections are **between the main breaker and the meter**. A feed-in tariff PV system (rarer in the US) would also be connected on the supply side of both the main breaker and the service meter. This type of installation could also be covered by 705.11.

We treat a source connection at a service, **like** a service, so we will mostly be using the rules from Article 230 Services, so **remember, Article 230 is Services**. We will be talking a lot about 230 here. Here is the Article 100 **definition for Service**: "The conductors and equipment connecting the serving utility to the wiring system of the premises served."

705.11(A) Service Connections

An electric power production source (usually PV or energy storage) can be connected to a service by one of the following three methods in 705.11(A)(1) through (3) and must comply with **705.11(B) through (F)** (pages 198–203).

705.11(A)(1) New Service

In accordance with 230.2(A), which is in **230.2 Number of Services** and just tells us that additional services can supply **interconnected power production sources**, as a special condition, among other things, such as fire pumps. Although the wording here says supply, isn't it we who are supplying them?

705.11(A)(2) Supply Side of Service Disconnecting Means

In accordance with 230.82(6), which is part of **230.82 Equipment Connected to the Supply Side of Service Disconnect**, much like with 705.11(A)(1), which is for new services, it tells us we can connect our power production equipment to the supply side of the disconnecting means for our **existing services** using overcurrent protection as specified in Part VII of Article 230.

Article 230 Services / Part VII Service Entrance Conductors is telling us to treat our supply-side connected systems like a service. This contains the wiring methods for services.

Take note that **230.46 Spliced and Tapped Conductors** tells us that we need to use pressure connectors and splices that are marked

"suitable for use on the line side of service equipment or equivalent." Check your Insulation Piercing Connectors (IPCs), if you are using them, to make sure they are suitable and can handle short circuits.

Even though we are directed to Article 230 Services, **solar and storage is not a service**, but is treated like a service when we look up our wiring methods. (We are talking about being an electrical service here and not how great of a service you offer your customers, and your solar service agreement.) The service we are talking about here is where the utility serves you electricity.

705.11(A)(3) Additional Set of Service Entrance Conductors

We can connect to an additional set of service entrance conductors in accordance with 230.40 **Exception Number 5**.

230.40 Number of Service-Entrance Conductor Sets has a limit of one set with many exceptions. The exception we are looking for is No. 5.

This exception tells us that we **can** connect systems covered by 230.82(5) and 230.82(6). We just covered 230.82(6), which tells us **we can connect our power source** on the supply side (page 197), and 230.82(5) tells us that we can connect **energy management systems (EMS), stand-by systems**, fire pumps, etc.

> **Microgrid Interconnect Devices and Source Connections at a Service (MID and microgrid defined on page 267)**
>
> Is it a **source connection to a service** (supply-side connection) when you connect a MID to service entrance conductors? Your MID will have a "main breaker" inside of it, so **the answer is no** since your power production source will be on the load side of that MID's service disconnecting means (main breaker).

705.11(B) Conductors (Source Connections to a Service / Supply-Side)

In the 2020 NEC, 705.11(A) was Output Rating and is now moved to be part of 705.11(B) Conductors as 705.11(B)(1)

705.11(B)(1) Output Rating (Ampacity ≥ Sum of Maximum Circuit Current)

The **sum of the power source continuous currents cannot exceed the ampacity of the service conductors**. The power source continuous currents are defined **in 705.28(A) Power Source Maximum Output Current** (page 205) and are just the continuous current ratings on the label of the interactive inverter or as programmed into an Energy Management System.

We do not take the size of the inverter OCPD, loads, or the size of the main breaker into consideration when determining the maximum amount of inverter rated current that we can connect with a source connection at a service. Typically, with a **source connection to a service**, you can connect as much PV as you will ever need. However, some large commercial facilities may want to install PV systems even larger than their service ratings to zero out their relatively high day and night electricity consumption. These large systems typically require a significant service upgrade, including a larger service transformer, or invoking 705.13 Energy Management Systems (EMS) [formerly Power Control Systems (PCS)]. We will learn about these systems soon on page 222 and how they **can control the power direction and quantity, just like we need it**. With the EMS/PCS, in theory, we can connect as much inverter capacity as we want, if our artificial intelligence (AI) pays attention to the current on the service conductors and makes sure there are not dangerous situations. This is the future! Even a current-limiting function on an inverter with a CT (current transformer) monitoring service conductors will perform this function.

705.11(B)(2) ≥ 6 AWG Copper or 4 AWG Aluminum

The **power source output circuit (inverter output circuit)** conductors from the point of interconnection to the first overcurrent device shall be sized in accordance with:

- **Source connections at a service** shall be at least **6 AWG copper or 4 AWG aluminum.**
- **705.28 Circuit Sizing and Current** (see Chapter 12 Wire Sizing). This is practically the same wire sizing techniques for all conductors carrying continuous current in the NEC. Nothing new here except organization.

705.11(B)(3) Ampacity of Other Service Conductors ≥ (B)(1) and (B)(2)

This is obvious and almost does not need to be stated. We need to comply with 705.11(B), which we are finishing now.

705.11(C) Connections (*Source Connections to a Service*)

The material here was organized into the list format, which is a theme for the 2023 NEC, more lists!

705.11(C)(1) Splices or Taps

Splices or taps shall be made in accordance with **230.33** for underground or **230.46** not underground.

230.33 Spliced Conductors is in **230 Part IV Underground Service Conductors** and sends us to 110.4, 230.46, 300.5(E), 300.13, and 300.15. We are going to list some of the places that we are sent to and name the different Sections; however, if we went into detail of every place we were sent, we would go in circles and this book would be too heavy to carry. We recommend looking it up in the NEC if you are doing a really weird **source connection to a service** or consulting with a really smart PE.

The connections shall be made using listed conductors as described in **110.14 Electrical Connections**. 110.14 pretty much says that we need to torque and install the listed conductors according to instructions, which was how they were tested during the listing process, while also paying attention to **110.14(C) Temperature Limitations**, which we will take into account in our wire-sizing chapter beginning on page 287 of this book.

230.46 Spliced and Tapped Conductors
300.5(E) Splices and Taps

110.14(B) Splices

300.13 Mechanical and Electrical Continuity – Conductors
300.15 Boxes, Conduit Bodies, or Fittings – Where Required
230.46 Spliced and Tapped Conductors is in **Part IV Service Entrance Conductors** and sends us to 110.14, 300.5(E), 300.13, and 300.15, which we just covered when discussing underground source connections to a service.

Additionally, the following devices must be listed as suitable for line side of service equipment (or equivalent):

- power distribution blocks
- pressure connectors
- devices for splices or taps

So, you cannot make a source connection at a service with something that is not service rated. We have to treat it like a service and remember that there can be higher currents coming from those service conductors.

705.11(C)(2) Existing Equipment

Modifications to existing equipment must be made in accordance with manufacturer's instructions, or field labeled for the application.

Discussion: we have seen many people do a source connection to a service (formerly supply-side connection) inside of all-in-one meter-main combo service equipment, which is not supposed to be done, unless it is able to be done according to the manufacturer's instructions or if you paid an NRTL to field label it.

705.11(C)(3) Utility-Controlled Equipment (new name in 705.11, but obvious)

For meter socket enclosures or other equipment under control of the utility, only connections approved by the utility are allowed.

705.11(D) Service Disconnecting Means (Source Connections to a Service)

- All ungrounded conductors of a power production source must run through a disconnect so they can be opened.
- Look to 230 Parts VI and VII.
  - 230 Part VI Service Entrance Conductors
  - 230 Part VII Service Equipment

Once again, we treat solar and storage (power production sources) like a service when connected on the supply side of the main service disconnect.

705.11(E) Bonding and Grounding (Source Connections to a Service)

Remember that when we are talking about source connections to a service, we are talking about alternating current and not that weird PV dc grounding.

Metal enclosures, metal wiring methods, and metal parts associated with a service connected to a power production source shall be bonded in accordance with **Article 250 Grounding and Bonding** Parts II through V and VII.

- 250 Part II System Grounding
  - This is where it tells us to ground a source connection at a service like a service and not to bond neutral to ground on the load side of the service equipment, like we used to in the exciting and sparky past.

> **Why We Bond a Source Connection to a Service "Like" a Service...**
>
> Some people worry about objectionable currents sneaking by, just like you cannot have two points of system grounding on a normal single service, such as bonding neutral to ground at a subpanel. The reason we treat a **source connection to a service**, like a service, is because we are on the supply side of all overcurrent protection, just like if you had a duplex and a separate meter and two separate service disconnects, then you have two separate points of system grounding. It is a big difference if you are on the supply side of all overcurrent protection. One of the problems with grounding and bonding, is that there is no perfect solution, and we have to look for the best solution, which sometimes can be different in different places, because of lightning, the conductivity of earth, and obviously crop circle vortexes. They probably do it this way on different planets since it is based on physics.

705.11(F) Overcurrent Protection (Source Connection to a Service)

In the 2020 NEC, we were directed to 705.30 Overcurrent protection; however, **now we are directed to Article 230 Services**, more specifically 230 Part VII and 230.95 (assuming ground fault protection is required).

- 230 Part VII Service Equipment Overcurrent Protection. Some of the more notable things here are:
  - Requiring OCPD in ungrounded conductors
  - OCPD part of service disconnecting means or immediately adjacent
  - Disconnecting means on the line side of fuses
- 230.95 Ground Fault Protection of Equipment (Included in Part VII) abbreviated
  - Ground-fault protection required for solidly grounded WYE services more than 150V to ground (not residential) and less than 1000V phase to phase (not typical).
- 230.95(A) Setting
  - Ground-fault protection shall open all ungrounded conductors of faulted circuit.
- 230.95(B) Fuses
  - If switch/fuse combination used, fuse shall be able to interrupt current higher than the interrupting capacity of the switch during time ground-fault protection system will not cause switch to open.
- 230.95(C) Performance Testing
  - Ground-fault protection shall be performance tested when first installed.

705.12 Load-Side Source Connections

Most solar installers and electricians prefer a load-side connection, since it is easy to turn off the main and safely pop in a solar breaker, just like they pop in a load breaker. Now with energy storage being so popular and expected to become more popular, it makes a load-side connection even more desirable. When you want backup power, you need to isolate from the grid and be on the load side of the service disconnecting means.

We are **permitted to connect** solar on the load side of ***any*** **distribution equipment on the premises.**

Examples of distribution equipment eligible for a load-side connection include:

- **Panelboards**

Article 100 Definition:

> A single panel or group of panel units designed for assembly in the form of a single panel, including buses and automatic overcurrent devices, and equipped with or without switches for the control of light, heat, or power circuits; designed to be placed in a cabinet, enclosure, or cutout box placed in or against a wall, partition, or other support; and accessible only from the front.

A main service panel and a subpanel are common examples of panelboards.

- **Switchgear**

Article 100 Definition:

> An assembly completely enclosed on all sides and top with sheet metal (except for ventilating openings and inspection windows) and containing primary power circuit switching, interrupting devices, or both, with buses and connections. The assembly may include control and auxiliary devices. Access to the interior of the enclosure is provided by doors, removable covers or both.

A switchgear acts like a panelboard for a big service.

- **Switchboards**

Article 110 Definition:

> A large single panel, frame, or assembly of panels on which are mounted on the face, back, or both, switches, overcurrent and other protective devices, buses, and usually instruments.
>
> Informational Note: These assemblies are generally accessible from the rear as well as from the front and are not intended to be installed in cabinets.
>
> Switchboards (to differentiate from switchgear) can meter, reroute power and divide incoming power into smaller circuits. **Switchboards are used at building electrical voltages whereas switchgear can be used at all voltages.**

Much of the load-side connection material in the 2023 NEC was rearranged, but it is mostly the **same in practice and theory** as it was in the 2020 NEC.

In the 2023 NEC, 705.12(A) Dedicated Overcurrent and Disconnect was changed to **705.12(A) Feeders and Feeder Taps**. Also, NEC 705.12(B) Bus or Conductor Ampere Rating was changed to **705.12(B) Busbars**, which formerly included the material which is now in 705.12(A) and the material still in 705.12(B).

2020 NEC 705.12(C) Marking, 705.12(D) Suitable for Backfeed, and 705.12(E) Fastening were moved from 705.12 to 705.30, so now we only have 705.12(A) Feeders and Feeder Taps, and (B) Busbars to keep track of in this section.

705.12 Load-Side Source Connections also tells us to comply with "relevant" sections of 705.12(A) and (B), so perhaps it is obvious to the AHJ what relevant is, at least after reading this chapter. Perhaps the next edition of the NEC should have an Informational Note telling us to read this book. Bill, can you set that up?

705.12 then tells us that the currents we use here are the maximum circuit currents calculated in **705.28(A) Power Source Output Maximum Current**, which essentially tells us to use the continuous current of a circuit that is limited by an energy management system (EMS), which we will get into when we study 705.13 Energy Management Systems.

705.12 then tells us to use 705.12(A) for feeders and taps and 705.12(B) for busbars.

705.12(A) Feeders and Feeder Taps

An example of a **feeder is a conductor that is going from a main service panel to a subpanel.**

705.12(A)(1) [Feeder Ampacity ≥ 125% Inverter Current] is so obvious that we do not need to study it. 705.12(A)(1) just tells us that the feeder ampacity must be at least 125% of inverter current, which is entry level wire sizing.

If we are going to connect to the middle of a feeder, we need to make sure that the conductor is properly protected on the **load side of the source connection**, and we apply **705.12(A)(2) [Load side of Inverter Connection to Feeder]**. The reason **we are concerned about**

the load side of the connection to the feeder rather than the supply side, is because by adding more supply current with our interactive inverter to the feeder, we are **no longer protected by the feeder supply breaker**. Before the interactive inverter was added, the feeder supply breaker was the protection for the feeder. However, by adding another current source, we can have the potential for overcurrents if we do not comply with **705.12(A)(2)(a) [Feeder Ampacity ≥ Feeder OCPD + 125% Inverter Current]** or **705.12(A)(2)(b) [OCPD ≤ Feeder Ampacity on Load Side of Connection Point]**.

705.12(A)(2)(a) [FEEDER AMPACITY ≥ FEEDER OCPD + 125% INVERTER CURRENT]

We add the feeder supply breaker plus 125% of the inverter current for this calculation. If the ampacity of the feeder on the **load side of the inverter connection** can handle the sum of these currents, then the feeder will be safe. If the existing feeder is not large enough, then **we can** replace that load-side portion of the feeder with a larger feeder. This would rarely ever happen in the field, but the option exists for those who want to use it.

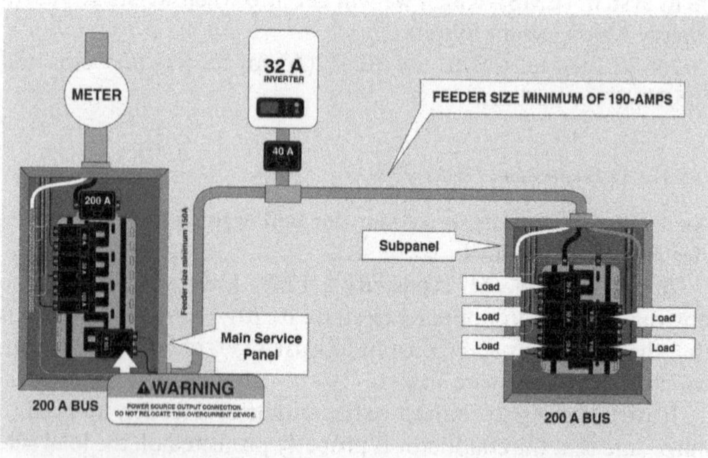

Figure 9.3 705.12(A)(2)(a) sufficient feeder ampacity.
Source: Courtesy Kylie Kwiatkowski.

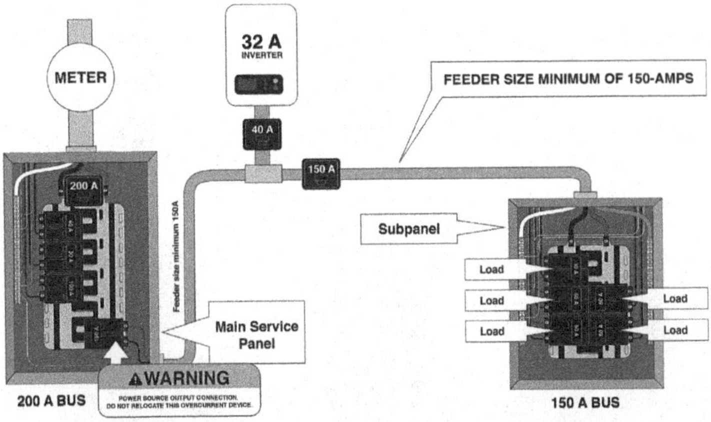

Figure 9.4 705.12(A)(2)(b) overcurrent device protecting feeder.

Note: If 25-foot tap rule is applied here, then the OCPD can be located up to 25 ft from inverter–feeder connection point, and perhaps in the subpanel, if the tap rules are followed, including the feeder being inside of a metal raceway.

Source: Courtesy Kylie Kwiatkowski.

705.12(A)(2)(b) [OCPD ≤ FEEDER AMPACITY ON LOAD SIDE OF CONNECTION POINT]

Another option rather than 705.12(A)(2)(a) is to place an overcurrent protection device on the load side of the connection of the interactive inverter to the feeder that is not greater than the ampacity of the feeder.

If we had a 100A feeder, a 100A feeder breaker, and a 30A inverter, we could place a 100A breaker on the load side of the connection between the interactive inverter connection to the feeder and the loads.

There have been different interpretations on where this breaker can be placed. The safest place to put the breaker is adjacent to the connection of the inverter circuit to the feeder. This was the intent when this provision was drafted by Bill in the 2014 NEC. Others have the opinion that a "main" breaker in the subpanel will provide this protection. The main reason for putting the breaker adjacent to the PV connection is that no one can argue that taps could be installed between the PV connection and the subpanel that could overcurrent the feeder conductor downstream of the PV connection.

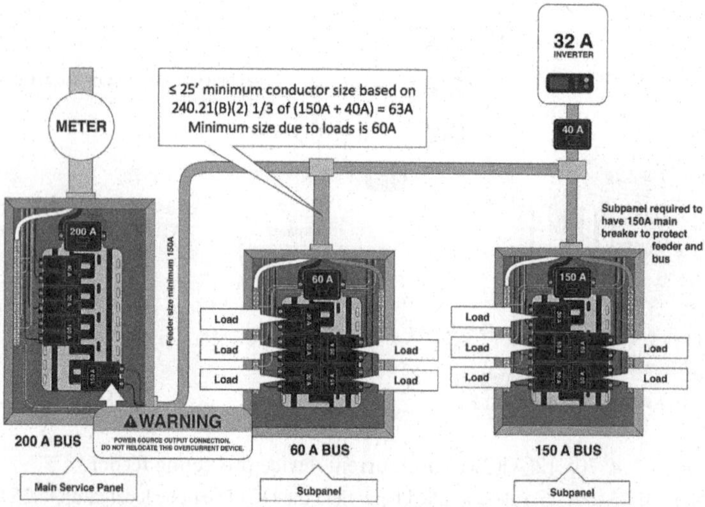

Figure 9.5 Solar tap rules.
Source: Courtesy Kylie Kwiatkowski.

We can best describe how far we can put the ***705.12(A)(2)(b) feeder overcurrent device*** from the connection point to the feeder by using the **705.12(A)(3) tap rules** which we will study in more depth in **705.12(A)(3) [Tap Conductors use 1/3 of feeder breaker + 125% Inverter Current]** coming up next. In most cases, we would use the **25-foot tap rule** and we can place the ***705.12(A)(2)(b) feeder OCPD*** in the subpanel if it is within 25 feet of the connection point to the feeder. We could also place that ***705.12(A)(2)(b) feeder OCPD*** on that conductor **anywhere within 25 feet of the connection point** (it does not have to be at the subpanel). This special case of putting the OCPD in the subpanel, rather than directly on the load side of the PV connection, assumes the **tap rules in 240.21(B)** (see page 208) are followed, which requires that the tap conductor is in a raceway and not an existing feeder that is not in a raceway, which is another reason to put the OCPD at the connection point. Few houses have conduit on any conductors other than their service conductors. There are some who think that they can place the ***705.12(A)(2)(b) feeder OCPD*** as far as they want from the connection point, and it is up to the AHJ to decide.

705.12(A)(3) [Tap Conductors Use 1/3 of Feeder Breaker + 125% Inverter Current]

Rather than referring to **240.21(B) Feeder Taps** in general (otherwise known as the "tap rules"), the **2023 NEC more specifically directs us** to **240.21(B)(2) Taps Not Over 25 ft (known as the 25-ft Tap Rule)** and **240.21(B)(4) Taps Over 25 ft**. This is **skipping the 10-foot tap rule since only those taps sized under the 25 ft tap rules are necessary to consider when applying 705.12(A)**.

Solar Tap Rules, Merging 705.12(B)(2) with 240.21(B)(2) and (B)(4)

This is complicated, so pay careful attention and read this a few times. Speed reading not allowed here.

We are connecting an extra current source to a feeder, so we need to make sure if we go a long distance that all conductors involved can handle a short circuit and in turn, open (turn off) that **feeder-protecting OCPD**, which supplies the line side of the feeder. We are not just looking at the inverter output circuit, we are also looking at the load side of the point of connection to any overcurrent protection device, so going from the **point of connection to the feeder to the overcurrent protection device on the load side connection point**.

If we have an inverter connected to a feeder, we should check the inverter output circuit. If we were applying the 25-foot Tap Rule for instance, we would need to make sure that there is an **OCPD within 25 feet of where the inverter is connected to the feeder.**

Electricians are familiar with the **tap rules in 240.21(B)(2) and (4)** and solar installers are often confused about what a "tap" is since as solar installers we are often and, for the most part, incorrectly calling a **source connection to a service** a "line-side tap."

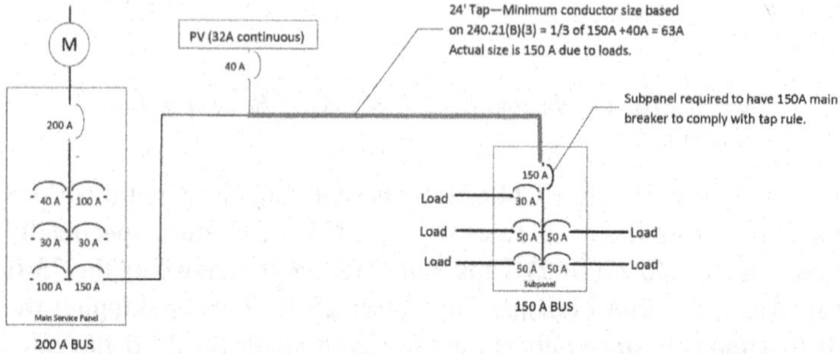

Figure 9.6 25-foot tap rule.
Source: Bill Brooks.

In order to apply the **tap rules in 240.21(B)(2) and (4),** we need an overcurrent protection device protecting the feeder that we are tapping into. A **705.11 source connection to a service** does not have any overcurrent protection on the supply side of the connection, thus **cannot follow the tap rules,** since there would be nothing to base your calculations on, with no feeder supply breaker.

The **705.12(A)(3) tap rules will add the feeder supply breaker to 125% of the inverter current** for this calculation.

25-foot tap rule for solar: If the conductor from the connection point to the OCPD is **less than 25 feet**, then the ampacity of that conductor between the connection point and the OCPD can be no less than **1/3 of the feeder supply breaker plus 125% of the inverter current**.

$$\text{Conductor to OCPD} \geq 0.33 \times [(\text{inverter current} \times 1.25) + \text{feeder supply breaker}]$$

The reason that we cannot go over 25 feet, is because a long wire has more resistance and is less likely to open up the feeder supply breaker in case of a fault.

Recall that we are sizing the conductor going from the point of connection to the OCPD here. We still would need to apply the 705.12(A)(2) feeder rules to make sure that the feeder itself is protected on the load side of the connection point. This is because applying the feeder rules and the tap rules together is "relevant."

The examples for the tap rules are related to PV and energy storage connections in this book. However, the very same rules are used for existing or new load taps and the required size of conductors for those taps—there is just no 125% of power source current without a power source.

705.12(B) Busbars

Every solar installer's favorite way to install solar is on a busbar, since popping in a breaker on a busbar is often the safest and easiest way of installing solar. Since we have currents coming from different sources on the busbar, we can be creative where we place the solar breakers and often get more out of the busbar than we would think at first glance.

We are about to explain how to do some math regarding how much current we can backfeed a busbar on a load-side connection. In the 2011 and earlier versions of the NEC, we were taught to use the inverter backfeed breaker size for our calculations. After the 2014 NEC we switched over to use 125% of the inverter current in most of our calculations. In many cases there will be no difference, but here are a few examples of how 125% of inverter current being used in the calculation can be beneficial:

- Rounding up to the next common breaker size.
  - Example: If we have a 3kW/240V inverter operating at 12.5A, we then multiply 12.5A × 1.25 = 15.6A. Since there are no 15.6A breakers, we then round up to a 20A breaker. If we use the 125% of inverter current in our calculation rather than the breaker size, we then get an extra 20A – 15.6A = 4.4A to play with.
- Using a 30A breaker, since a 25A breaker is uncommon.
  - Example: If we are using a 4kW inverter at 240V then our inverter current is 4kW/240V = 16.67A and 16.67A × 1.25 = 20.8A and in this case we can round up to a 25A breaker; however, electricians may find it difficult to locate a 25A breaker and often use a 30A breaker. It is acceptable

to use a 30A breaker in this case as long as the conductor is large enough to be protected by a 30A breaker and as long as the inverter manufacturer allows a 30A breaker to protect the inverter. In this example 125% of inverter current is 20.8A and the breaker is 30A, so we have 30A–20.8A = 9.2A more of an allowance by using the 125% of inverter current rather than the breaker size method.

- Having a few microinverters on a circuit. Often with microinverters we do not have the maximum number of inverters on a circuit. At times when the microinverter circuit has a long way to go to reach the interconnection, the designer will place fewer microinverters on a circuit to address voltage drop considerations. Other times, we see a few microinverters on a circuit, just because it is what fits on the roof.
    - Example: If we have three 250W microinverters on a 20A breaker at a house, then inverter current would be calculated 250W/240V = 1.04A and 125% of the current for each inverter would be 1.04A × 1.25 = 1.3A and for three inverters 125% of current is 1.3A × 3 = 3.9A. In this case the benefit of using 125% of inverter current rather than the breaker size is 20A – 3.9A = 16.1A benefit, so in this case it can make a big difference.

125% of inverter current is more difficult to explain, but is worth it, since it allows for more PV to be installed than in previous Code cycles. When I do explain it, I often say "the largest possible backfeed breaker" instead of "the backfeed breaker," as I used to in 2011.

705.12(B)(1) through (6) are not named and are options and not rules (at least when "relevant"); however, since it is so common to call 705.12(B)(2) the 120% rule, we will go ahead and let you call it the 120% rule. We tried to make everyone change "rule" to "option," but nobody complied, including ourselves.

Some new wording for 705.12(B) Busbars goes:

> For power source connections to distribution equipment with **no specific listing** and instructions for combining multiple sources, one of the following methods shall be used to determine the required ampere ratings of the busbars.

Take note that it says "**with no specific listing and instructions for combining multiple sources.**" This means that if you did have special listed equipment for combining multiple sources, then you do not have to comply with these 705.12(B) rules (options). Perhaps your inverter manufacturer, busbar broker, or panelboard pusher has a special listed busbar with instructions how you can have a 200% rule. Go for it!

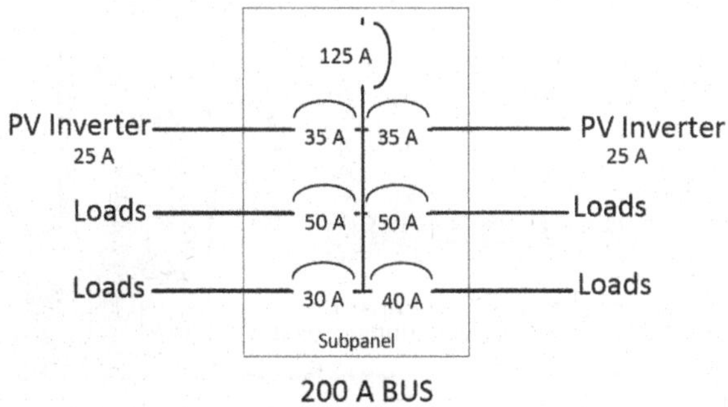

Figure 9.7 100% option.
Source: Bill Brooks.

705.12(B)(1) 100% Rule (Option)

If 125% of the inverter current plus the main breaker does not exceed the rating of the busbar, we can place the inverter breaker anywhere on the busbar (it does not have to be on the opposite end from the main supply breaker).

For example, if we have a 200A busbar with a 125A main breaker, we can place two breakers for two 25A inverters anywhere we want on the busbar. We can place up to 75A / 1.25 = 60A of inverters anywhere we want on the busbar. We can also do the inverse math 75A × 0.8 = 60A. 0.8 is the inverse of 1.25. (The 75A comes from subtracting the 125A main breaker from the 200A busbar.)

Some installers think that the inverter breaker always must go on the opposite side of the busbar from the main breaker. **This is not true** if 125% of the inverter current plus the main breaker does not exceed the rating of the busbar. The 100% rule is just an option.

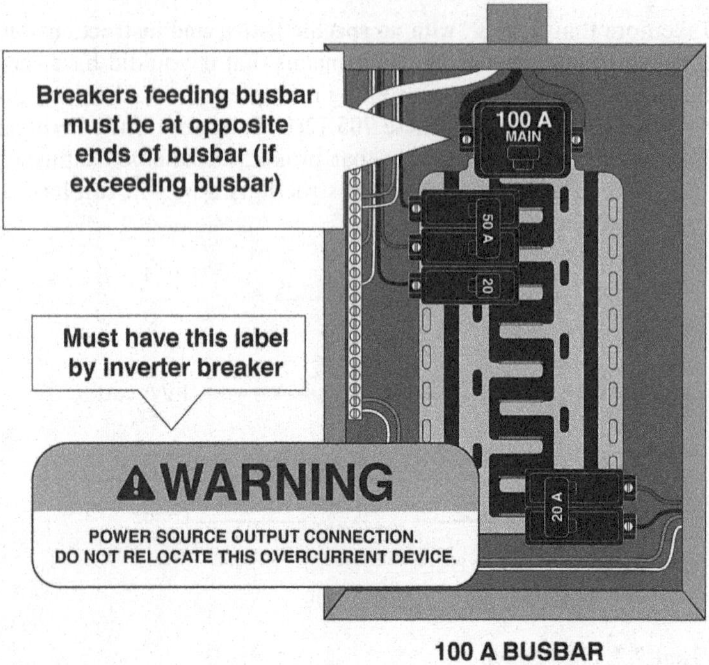

Figure 9.8 705.12(B)(2) 120% option (rule).
Source: Courtesy Kylie Kwiatkowski.

705.12(B)(2) 120% Option (Formerly Known as 120% Rule)

NOTE: **We changed the term from 120% "Rule" to 120% "Option"** in the last version of this book to make a point, so we might as well stick with it. This is because so many installers are fixated on the 120% restrictions, that it prevents them from using some of the better options that have been available since the 2014 NEC.

$$\text{Busbar} \times 1.2 \geq \text{Main Breaker} + (1.25 \times \text{inverter current})$$

We can exceed the rating of the busbar by up to 20% after adding the main supply breaker plus 125% of the inverter current if the supply breaker and the solar breaker are on opposite ends of the busbar. This opposite end clause was once interpreted in such a way that denied center-fed panelboards from being able to apply the **120% option**, but as we will soon see in 705.12(B)(4), this is no longer the case for dwellings.

Here is the **120% option math** from a few different angles:

Main + (1.25 × Inv current) ≤ busbar × 1.2

1.25 × Inv current ≤ (busbar × 1.2) – main

Inv current ≤ ((busbar × 1.2) – main)/1.25 or

Inv current ≤ ((busbar × 1.2) – main) × 0.8 (note: 0.8 = 1/1.25)

Maximum inverter power formula using **120% option**:

grid voltage × (((1.2 × busbar) – main) × 0.8) = max inverter power

Recall that backfed breakers will have to be located on the **opposite side of the busbar** from the main breaker and that there shall be a label saying the following words **or equivalent**:

WARNING:
POWER SOURCE OUTPUT DO NOT RELOCATE THIS OVERCURRENT DEVICE

The reason that we can exceed the busbar rating is because we have currents feeding the busbar coming from different directions, which makes it easier on the busbar and prevents busbar "hot-spots" as could happen if the backfeed breaker were put next to the main breaker. The **120% option** will take some heat off the main breaker when power is fed from a spot on the busbar distant from the main breaker.

Another thing that people sometimes get confused about is thinking that they can only have one backfeed breaker at the opposite end. You can have multiple breakers. If you have four backfeed breakers, then just make sure they are at the four farthest spots from the main breaker. One fun way to exploit the 120%

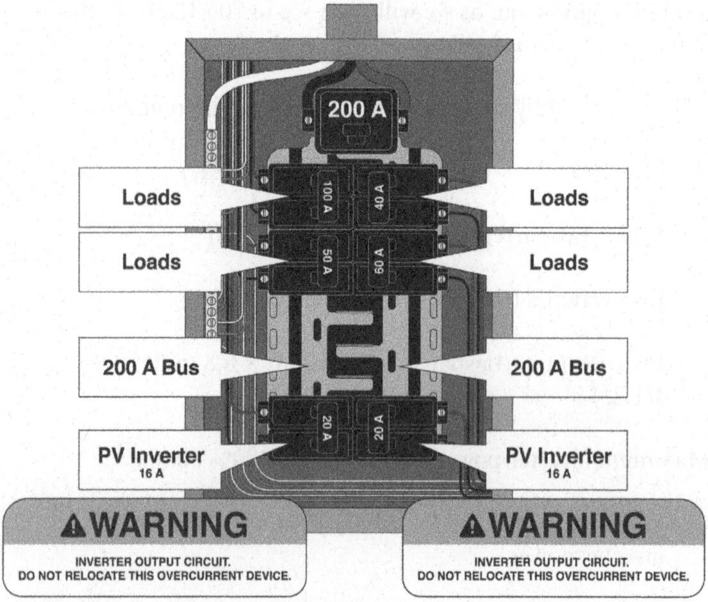

Figure 9.9 120% rule (option) with multiple solar breakers acceptable. *Source:* Courtesy Kylie Kwiatkowski.

rule (option) is to replace your 100A panelboard with a 200A panelboard, but keep the 100A main breaker. Then you can add a bunch of bidirectional EV breakers and make a lot of money selling grid support services to the utility via a virtual power plant (VPP) in the near future (we know the future).

705.12(B)(3) Sum of the Branch Breakers Option a.k.a. Sum Rule (See figure 9.10 next page)

This option was primarily created so that subpanels could be logically used for **ac combiner** panels **without the restrictions of the 120% option**. This is a very simple option that uses the sum of the branch circuit breakers in a panel to protect the busbar of the panel. These branch circuit breakers can be any combination of generation and load breakers. Take note here that this is the only load-side connection that uses the ratings of the circuit breakers and not 125% of inverter current in the calculations.

The easiest way to understand how this rule works is by reading the label required to be installed on the distribution equipment which reads:

> Warning:
> This Equipment Fed by Multiple Sources.
> Total Rating of All Overcurrent Devices
> Excluding Main Supply Overcurrent Device
> Shall Not Exceed Ampacity of The Busbar.

In effect, what this "**Sum Thing/Rule/Option**" is doing is protecting the busbar in reverse, by making sure that the **sum** of all the breakers on the load side of the busbar protect the busbar.

If you looked in the main service panel in your house, you would likely find that the sum of the branch circuit breakers is much more than the main breaker of the busbar is rated for. We just do not turn on everything at once, so the main breaker does not trip.

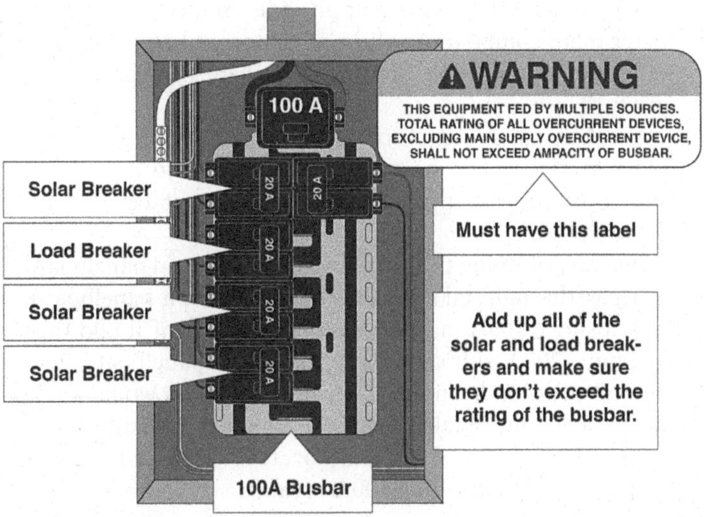

Figure 9.10 705.12(B)(3) sum option.
Source: Courtesy Kylie Kwiatkowski.

With the "Sum Rule" we are protecting the busbar by making sure that we do not have too much current going from the inverter breakers towards the busbar. This rule is conservative if we have loads on the busbar.

One application for someone trying to avoid a supply-side connection (which is really good for energy storage with backup) is to put loads from a main service panel onto a subpanel until you have enough space on your main service panel to apply the "Sum Option" and add more solar. In Hawaii they are famous for getting creative with the Sum Option and they call it the **Hawaiian Tie-In**! Imagine taking a few subpanels and connecting them to an all-in-one meter main and applying the Sum Rule this way.

Hawaiian Tie-In (See figures 9.11 and 9.12 next page)

There have been some creative and much needed installations done connecting to a feeder or using the Sum Option and we would like to hand it to the islanders for coming up with these. The problem is that it is difficult to make a source connection to a service (supply-side connection) with a meter-main combo or an all-in-one meter main panelboard, so instead we can do what you see here in Figures 9.11 and 9.12.

Another good aspect of using this method is, as we add batteries, it can be more advantageous to do a load-side connection. Supply-side batteries cannot be used for backup power.

One way of doing this is visualizing your main panelboard as a passthrough: take the existing 100A main panelboard, stick a single 100A main breaker coming out of it and then feeding a 200A subpanel with that 100A panelboard, and have all your loads on the 200A subpanel. You would apply the Sum of the Breakers Option rule at the existing main panelboard and the 120% rule or 100% rule at the 200A subpanel being fed with a 100A breaker.

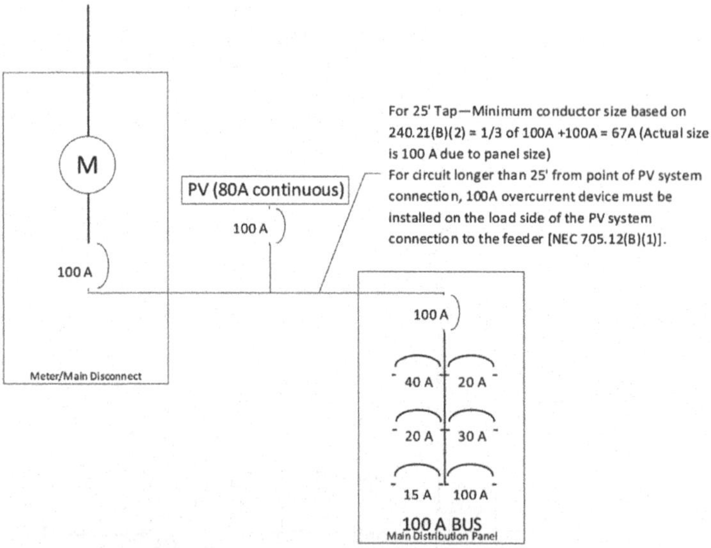

Figure 9.11 Hawaiian tie-in option using sum option.
Source: Courtesy Bill Brooks.

705.12(B)(4) Center-Fed 120% Option for Dwellings

Center-fed panelboards are main service panels or subpanels that are fed *not* from one end, but will have loads connected on both sides of the main breaker.

Millions of dollars were spent on upgrading main service panels before 2016 when we could not apply the 120% rule to center-fed panelboards, until there was a TIA (tentative interim amendment) in the summer of 2016 modifying the 2014 NEC to allow applying the 120% rule to center-fed panelboards.

This TIA means that if you are in some weird place that is still using the 2014 NEC, you can now apply the 120% option to center-fed panelboards on dwellings!

In the 2017 NEC, the proper way to apply the **120% rule to center-fed** panelboards is to **only connect solar PV to one side of the busbar and not both**. In the 2020 NEC, it was changed to be allowed on both ends! Total from both ends can still only be 120%—nice try, if you wanted to put 40-amp PV breakers at both ends of a

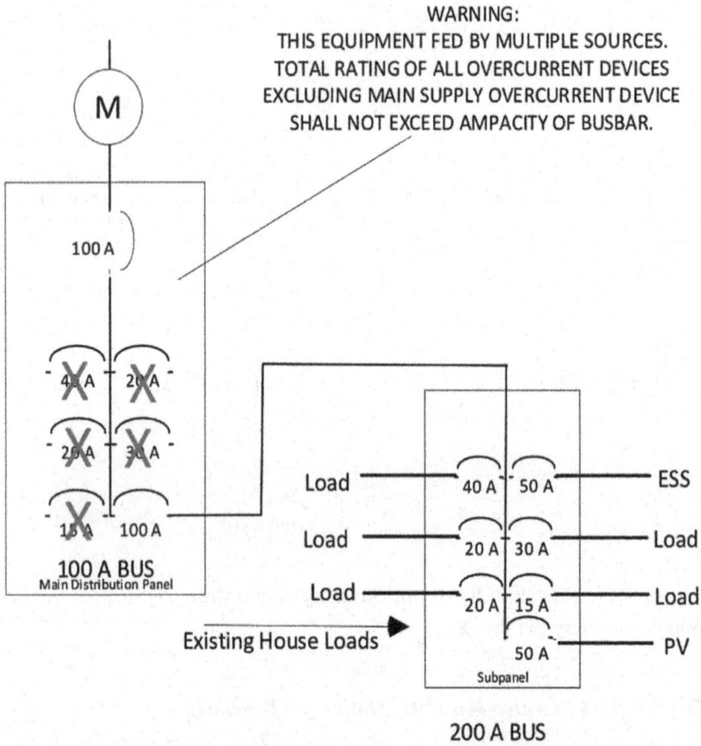

Figure 9.12 Hawaiian tie-in Mai Tai option using sum option.
Source: Courtesy Bill Brooks.

200-amp panel—that is not allowed. Stay tuned for 705.13—you will love it if you want to do the million % rule (you just have to be able to control your currents).

705.12(B)(5) Feed-Through Conductors/Lugs

Some panelboards have lugs at the bottom of the busbars. If we are feeding these feed-through lugs we shall treat the feed-through conductors coming off the feed-through lugs as if they were feeders according to **705.12(A) Feeders and Feeder Taps**. In this case, when applying 705.12(A), we will want to protect the **feed-through conductors/feeders** as feeders with extra currents that come

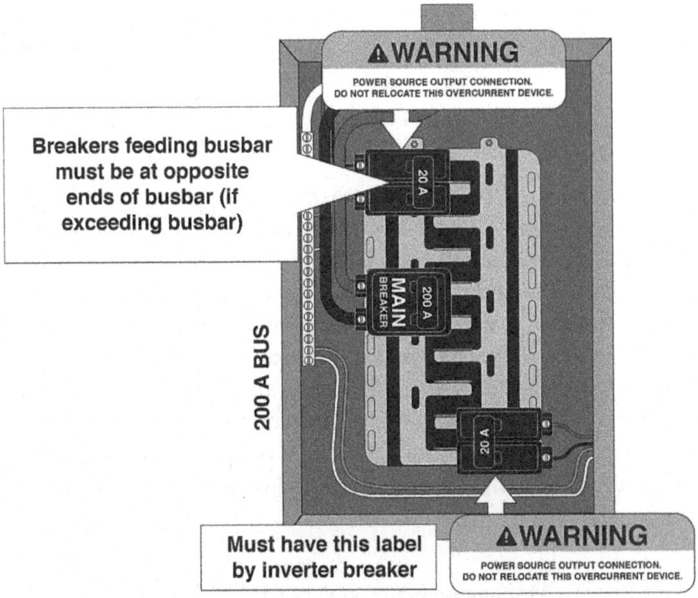

Figure 9.13 705.12(B)(4) center-fed 120% option for dwellings.
Source: Courtesy Kylie Kwiatkowski.

from 125% of inverter current plus the feeder supply breaker. In this case, the feeder supply breaker is the main breaker from the panelboard containing the feed-through lugs and this could end up causing the **feed-though conductors/feeders** to be sized rather large [1/3 of (main breaker + 125% inverter current)] and limiting you to 25 feet (or 100 feet in high bay buildings). In this case, we can look at the feed-through conductors to be an extension of the busbar.

Another option is putting overcurrent protection at the supply end of the feed-through conductors/feeders and using the **busbar methods in 705.12(B)(1) through 705.12(B)(3)**.

Recall:

705.12(B)(3)(1) is the 100% Option
705.12(B)(3)(2) is the 120% Option
705.12(B)(3)(3) is the Sum Option

Once we have overcurrent protection by the feed-through lugs, then we have a "supply breaker" to work from and applying these rules is simpler.

705.12(B)(6) Switchgear, Switchboards and Panelboards Under Engineering Supervision

According to the NEC, there are many things that we can do under **engineering supervision**, and connecting to switchgear, switchboards, and panelboards is one thing that we can accomplish.

To make these connections, which would typically be on commercial buildings, we have to have an **engineer supervise a study of available fault currents and perform busbar load calculations**.

This option is underutilized. There are many times when an engineer is already working on the project and can just sign off on something analogous to the 150% rule, or something else safe, if you get my drift. Engineers have special powers when it comes to sizing busbars.

705.13 Energy Management Systems (EMS) [Formerly Power Control Systems (PCS)]

Article 750 is Energy Management Systems and when the 2020 NEC was made, the stoic geniuses in charge of 750 did not want to be involved with these crazy currents going backwards, so we called it Power Control Systems in the 2020 NEC, but now there has been a change of heart and the 750 people do want to have power over us backwards current people, who, by the way, are taking over.

The idea with 705.13 in the 2020 and 2023 NEC is that we can control the flow of currents on busbars and conductors with smart listed electronics, rather than using the rules and options in 705.11 Source Connections to a Service and 705.12 Load-Side Source Connections.

705.13 Is the Future!

This is forward thinking and now there are over a dozen products that have been listed to the newer PCS requirements in UL 1741. With abundant PV, more EVs, and greater electricity usage, it is essential that we have means to control our currents and to use our huge, and currently underutilized, batteries with wheels to be able to support the grid. If you do the math, with a typical house that uses 10,000kWh per year, the average current is closer to 4A than 5A, although this same house is required to have a 200A service. In 10 years, when this house has two or three EVs with 250kWh of storage, 80% of which is not being utilized on a daily basis and the EVs have bidirectional capability, we can then start thinking about trickling closer to 5A than 20A from the utility. We can export to the utility when it gets hot, cold, or whatever the electrical system needs, and we are going to be using perhaps double the electricity that we use in 2023 when we move toward 2033. That means something like 9A continuous throughout the year, and we can and will easily do this with our 705.13 ingenuity. Besides bidirectional EVs, we will see virtual power plants (VPP), which are brokers for coordinating with customers and the utility while supporting the grid, especially with energy storage. Obviously, we will use much more distributed solar, energy storage without wheels, 2nd life EV batteries, smart load centers, thermal storage of heat and cold, smart appliances, and AI controls to tie it all together. We could not properly end this thought, without stating that silicon is smarter than us (AI).

2023 NEC 705.13 is one short sentence, which we will display here:

> An EMS in accordance with 750.30 shall be permitted to limit current and loading on busbars and conductors supplied by the output of one or more interconnected electric power production or energy storage sources.

Since 705.13 now sends us to 750.30, it is time to dive into the 750 EMS zone…

750 Energy Management Systems is a short article and contains **750.30 Load Management**, which is slightly misleading, since **solar and such is the anti-load!**

Here is a little outline of 750.30:

750.30 Load Management
- 750.30(A) Load Shedding and Controls
  - 750.30(A)(1)–(5) Lists loads to not shed, such as emergency and fire things
- 750.30(B) Disconnection of Power
  - 750.30(B)(1)–(5) Lists loads to not disconnect, such as elevators and essentials
- 750.30(C) Capacity of Branch Circuit, Feeder, or Service
  - 750.30(C)(1) Current Setpoint
    - 750.30(C)(1)(1) Calculation in 220.70 Energy Management Systems (EMSs) (new) (in Article 220 Branch-Circuit, Feeder, and Service Calculations)
    - 750.30(C)(1)(2) Maximum source current permitted by EMS
  - 750.30(C)(2) System Malfunction
  - 750.30(C)(3) Settings (at least one of the following)
    - 750.30(C)(3)(1) Located behind removable and sealable cover
    - 750.30(C)(3)(2) Located behind cover that requires tool
    - 750.30(C)(3)(3) Located behind locked door accessible to qualified personnel
    - 750.30(C)(3)(4) Password protected
    - 750.30(C)(3)(5) Software with password
  - 750.30(C)(4) Marking
    - 750.30(C)(4)(1) Max Current Setting
    - 750.30(C)(4)(2) Date of Calculation
    - 750.30(C)(4)(3) Identification of Loads and **Sources** with Current Limiting
    - 750.30(C)(4)(4) Wording or equivalent "The setting for the EMS current limiting feature shall not be bypassed"

As we can see from the outline, 750.30 is very load-centric; however, there are EMS rules that we must follow, such as current setpoints, making the EMS only programmable by qualified people. Marking and labeling requirements are obvious when reading the outline above. We do not want Joe homeowner coming along and increasing the current on his busbar, which can cause Joe firefighter to make a special trip to save his family.

Diagrams Using 705.13 Energy Management Systems

Here are a few ways to take advantage of this EMS piece of Code. See figures 9.14 and 9.15.

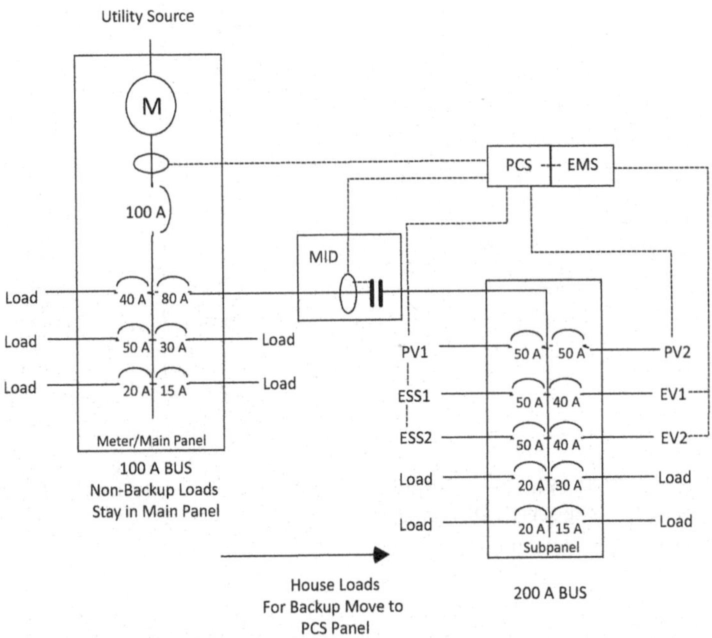

Figure 9.14 Backup power using microgrid interconnect device (MID) downstream of main.

Source: Courtesy Bill Brooks.

In a way, the 2023 NEC for **interconnected electric power production sources** is less specific on 705.13 than the 2020 NEC version, since in the 2020 NEC it was written by renewable energy people and now that 705.13 just refers us to 750.30 EMS, the focus of NEC 750 EMS is on loads. Whenever something new happens, it takes a while for the wrinkles to be ironed out, but the intention and idea are the same.

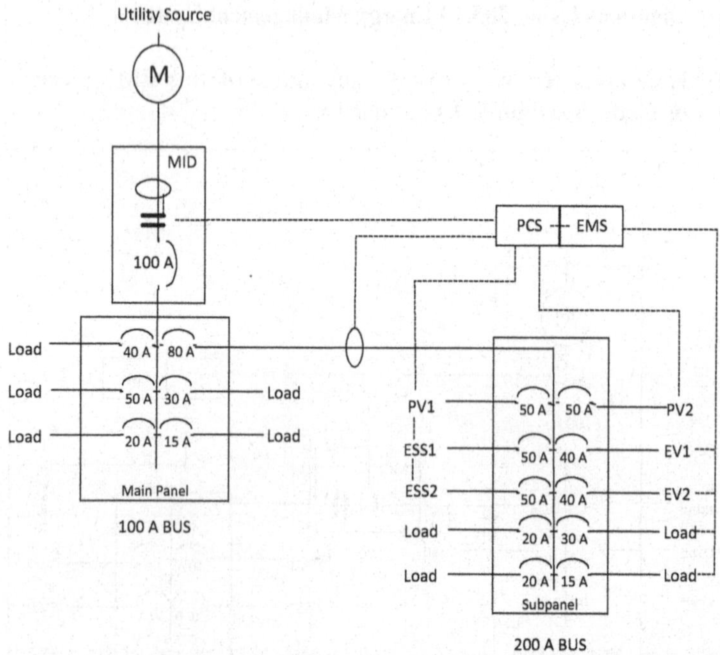

Figure 9.15 Backup power using microgrid interconnect device (MID) at the service disconnect.

Source: Courtesy Bill Brooks.

In the 2023 NEC, "Power Control Systems" are only mentioned twice in two Informational Notes, one in **705.6 Equipment Approval** and the other in 705.13 EMS. The **705.13 Informational Note** tells us that **a listed power control system (PCS) is a type of an EMS that can monitor multiple power sources and control currents on busbars and conductors**. The 705.6 Informational Note mentions that PCSs commonly have interactive functions.

An EMS needs to be listed and evaluated to control the output of power production sources, energy storage systems, and other equipment. The EMS will limit currents on busbars and conductors supplied by the EMS.

This means that we can use electronics (think of that trendy acronym AI) to control the currents flowing around the circuits and busbars. Now you can theoretically connect PV, an energy storage system (ESS), a bidirectional electric vehicle, and load controls so that you can supply as much as 400A while having a 100A service! Perhaps you need a fast charge on your new EV, so you take 200A from your ESS, 50A from your PV, 50A from your utility, and another 100A from your old EV to get 400A while only drawing 50A from the utility.

The 2020 NEC explicitly stated that on the load side of the service disconnecting means **any currents not monitored** by the PCS [EMS] shall comply with **705.12 Load-Side Source Connections**. This is no longer explicitly stated in the 2023 NEC but it is implied in the requirements of 705.28.

The 2020 NEC also stated that if the PCS [EMS] is connected on the supply side of the service disconnecting means as with **705.11 Source Connections**, then PCS [EMS] will monitor currents on service conductors to prevent overload of service conductors. This is also no longer explicitly stated but should be implied. We just need to use common sense here which is in very short supply these days with all the supply-chain issues.

This is a basic idea that ensures that the settings of the EMS are capable of monitoring and controlling the power sources to a subpanel or conductor within the current ratings or ampacity of the equipment. Where a EMS does not have control of all the sources feeding a subpanel (e.g., the utility feed), then the settings on the EMS must prevent more than the rated current to be fed to the subpanel. Therefore, if a 100-amp panel is being fed by PV and ESS via an EMS and the current begins to exceed the ratings of the 100-amp panel, then the EMS would have to reduce the PV and ESS power causing the 100-amp feeder breaker from the utility to trip. This scenario would actually only happen in an overloaded panel.

Another principle is that the EMS shall function as an OCPD or work with other OCPDs and the rating of an OCPD for a device for a single power source controlled by an EMS shall not exceed the rating of the busbar or conductor.

This principle means that, although an EMS may be controlling the current, we can never connect a single power source to a busbar that is larger than the busbar current rating.

750.30(C)(3) Settings is much like 240.6(C), which we used in the 2020 NEC.

In other words, when we are using an EMS to prevent overcurrents, we need to make sure that some burger flipper does not start messing with the settings to get the deep fryer working after it keeps tripping the breaker. An analogy would be how people in the 1970s used to put a penny under the fuse since the fuse kept blowing. It could get hot—and we are not talking 2°C (3.6°F) of global warming here, we are talking about flaming Venus 911 hot!

705.20 Source Disconnecting Means

Ungrounded conductors from power production sources must have a means for disconnecting. This applies to most anything in the NEC and was already covered in Section 690.13 on page 79.

One notable change in 705.20 is that in the 2023 NEC, it was clarified that a single disconnecting means shall be permitted to disconnect multiple power sources from conductors or other systems. Perhaps this could be the SolarEdge StorEdge disconnect that works for the PV array and the battery, all with one twist of a knob.

705.25 Wiring Methods

The wiring methods here are mostly duplicative of what we see in 690.31 Wiring Methods, which refers a lot to Chapter 3 Wiring Methods and Materials, and we have already covered this, beginning on page 100 of this book.

705.28 Circuit Sizing and Current

Circuit sizing and current for PV systems is covered in **690.8 Circuit Sizing and Current** and covered beginning on page 34 of this book. Additionally, you will find a lot of this information and how to use this information when we cover wire sizing in Chapter 12 of this book. Circuit sizing and current for PV systems and other loads or power sources throughout the Code follows the same rules and theory for the most part.

New in 2023 there is mention of EMS (Energy Management Systems) in 705.28 and these systems can control currents, so we use the "**current setpoint**" for our EMS controlled **current value.**

Most of the time an interactive inverter ac grounded conductor (neutral) does not carry current and is just there for

instrumentation, voltage detection, and phase detection. In this case, and going along with the Code, we use **Section 250.102 Grounded Conductor, Bonding Conductors, and Jumpers** to size the grounded conductor, which sends us to **Table 250.102(C)(1) Grounded Conductor, Main Bonding Jumper, System Bonding Jumper, and Supply-Side Bonding Jumper for Alternating Current Systems**. In this table we can **select the non-current-carrying grounded conductor based on the size of the current-carrying conductor**. For example, a **1/0 copper ungrounded conductor calls out for a 6 AWG (or larger) grounded conductor**. This is not special for renewable energy but is mentioned in **705.28(C)(2) Neutral Conductor Used Solely for Instrumentation, Voltage Detection, or Phase Detection**.

Some inverters, from small Enphase inverters to other very large inverters, do not even have an ac grounded conductor.

705.30 Overcurrent Protection

This lines up very closely with **690.9 Overcurrent Protection** on page 49 of this book and **Article 240 Overcurrent Protection**. You can also refer to Chapter 12 Wire Sizing in this book to see how to select overcurrent protection devices. 705.30 reminds us that **240.4(B) Overcurrent Devices Rated 800A or Less** allows us (with exceptions) to round up the OCPD to next higher ampacity above the conductor's ampacity if the resulting OCPD is below 800A. This is the step in wire sizing that is not completely logical at face value and we have an example in Chapter 12 where the ampacity of the OCPD is greater than the ampacity of the conductor that it is protecting on page 293. Having an OCPD with a greater ampacity than the conductor is not common.

240.4(C) Overcurrent Devices Rated over 800A tells us the opposite of 240.4(B), that we cannot have an overcurrent device that is greater than the conductor ampacity it is protecting.

Here is a summary of this section:

705.30(A) Circuit and Equipment

Power source output conductors shall have overcurrent protection from all sources.

705.30(B) Overcurrent Device Ratings

125% of maximum (continuous) current (typical throughout the NEC).

705.30(C) Marking (used to be in 705.12)

Equipment containing OCPD shall be marked to indicate power sources

705.30(D) Suitable for Backfeed (used to be in 705.12)

Fused disconnects (unless otherwise marked) are considered suitable for backfeed. Even circuit breakers marked line and load are suitable for backfeed if specifically rated.

705.30(E) Fastening (used to be in 705.12)

Interactive inverters do not require additional fastener normally required by 408.36(D) (which is required for stand-alone circuits).

705.30(F) Transformers

Primary side is considered the largest source of fault current (grid side). For more on transformers see page 60.

705.32 Ground-Fault Protection

Interactive systems shall be installed on the **supply side of ground-fault protection equipment**.

705.32 Exception (connecting to the load side of ground-fault protection)

If there is ground-fault protection from all ground-fault current sources, then it is acceptable to connect to the load side of ground-fault protection equipment.

Discussion: It is difficult to protect against ground-faults if there are current sources on each side of ground-fault protection. Many solar professionals do not bother trying to connect on the load side of ground-fault protection. All new services of 1000A and larger, built after 2002, are required to include ground-fault protection. Most manufacturers of these systems are accustomed to providing documentation that their detection circuits can handle backfed power from a generator (like an interactive inverter) on the load side of the ground-fault protector.

705.40 Loss of Primary Power Source

We are not allowed to feed the grid, when the grid (primary power source) is down!

Discussion: UL 1741 inverters are safe with regards to not feeding the grid when the grid is down, a.k.a., anti-islanding.

Since Article 705 includes other sources of power being connected to the grid, including big diesel generators, there is a provision in 705 to require conductors to be disconnected automatically from the primary power source.

Interactive inverters, however, will stay connected and monitor the grid-connection and export power five minutes after a good clean grid power source is indicated. Interactive inverters cannot feed the grid when the grid is down. No exceptions. It cannot happen and it is certified to prevent such an occurrence.

Multimodal inverters can power loads when the grid is down, and this is done with a separate output circuit coming out of the inverter that is not connected to the utility grid, or with a microgrid interconnect device (MID). Multimodal inverters can have multiple (usually two) outputs that operate in different modes, such as an interactive mode output and a stand-alone (island) mode output. Many newer multimode inverters have a single output and perform anti-islanding with a separate microgrid interconnect device (MID). See Island Mode Definition on page 266 in Chapter 11, or in Article 100 Definitions.

Chuck Norris Exception: When Chuck Norris's PV system disconnects from the grid, the grid goes into island mode. Source: www.chucknorrisfacts.net

Discussion: According to UL 1741, listed interactive inverters will stop exporting power when one phase is down.

Other than interactive inverters must disconnect all phases when a phase is down.

It is not mentioned here in 705.40, but **it is acceptable to have single-phase inverters connected to a 3-phase service** and if a phase is down that is not connected to a single-phase inverter, then the single-phase inverter is allowed to export power on the good phases.

Often single-phase microinverters are connected to 3-phase services via a cable that rotates the phases by connecting different inverters to different phases.

705.45 Unbalanced Interconnections

705.45(A) Single Phase

Single-phase inverters (and ac modules) shall be connected to 3-phase systems in order to **limit unbalanced voltages to not more than 3%.**

> **Single-Phase Inverters on 3-Phase Services**
>
> Just as loads can be strategically placed to limit imbalances on 3-phase busbars, so can single-phase inverters, but with reverse logic as with loads. If a building is out of balance and **inverters are placed on the phases with lower voltage,** which are the most heavily loaded phases, the inverters will reduce the utility load on those phases and help raise the voltage by sending power in a reverse direction. The result is that the imbalance of the service is reduced and the neutral current also goes down. Inverters cause voltage rise on a backfed busbar. Remember that with an energy storage system, your electronic power converter (inverter/charger) is not just a source, but also a load when charging, so it can cause voltage to rise and drop.
>
> It is conceivable that someone could invent an inverter that could balance the voltages of different phases by injecting power where there is a phase with less voltage. This could even happen at night.
>
> It is normal to have more loads on one phase than another and bringing along smart inverters can reduce this imbalance, so we can be the solution, not the problem.

705.45(B) Three Phase

- Three-phase inverters (and ac modules) shall have all phases de-energize upon unbalanced voltage in a single phase unless the inverter is designed so that significant unbalances will not result.

Discussion: UL 1741-listed inverters will automatically disconnect with large imbalances. There are new inverter standards, which have been adopted in California at the time of the writing of this book, such as UL 1741 SA, which will change the way inverters help support the utility grid, and other standards that may have been released while you were reading this book. These new developing standards address a new level of saturation of distributed generation on the grid and can differ based upon where you live. Exciting times!

Part II Microgrid Systems (Includes DC Microgrids in 2023 NEC)

Dc microgrids were formerly covered in Article 712 in the 2020 and 2017 NEC, but now they are all covered here. Not very many people were using Article 712 and the theories are the same for dc and ac in this case and most cases.

705.50 System Operation (Microgrid Systems)

Microgrids can operate in interactive mode with a primary power source and can disconnect from that primary power source (usually utility) and then operate in island mode.

Microgrids can include PV, wind, energy storage systems, and even internal combustion engine (ICE) generators (obviously powered with clean green recycled organic homegrown sustainably harvested virgin olive oil biodiesel).

705.60 Primary Power Source Connection (Microgrid Systems)

Connections to primary power sources (grid) shall comply with:

- **705.11 Supply-Side Source Connections (page 196)**

- **705.12 Load-Side Source Connections (page 203)**
- **705.13 Energy Management Systems (formerly Power Control Systems) (page 222)**

705.65 Reconnection to Primary Power Source (Microgrid Systems)

Reconnection shall be provided with necessary equipment to establish a synchronous connection. Interactive and multimode inverters all provide this service.

705.70 Microgrid Interconnect Devices (MID) (MID definition on page 267)

See page 267 of this book for the definition of MID. Many **multimode inverters** have always had a MID internally, which would disconnect from the grid and provide backup power to protected loads. Many newer MIDs are external to the inverter and signal the inverter to go into island mode when the grid goes down.

MIDs are/must:

(1) Required between microgrid system and primary power source
(2) Be listed or field labeled
(3) Have OCPD to protect from all sources

705.76 Microgrid Control Systems (MCS) (New in 2023 NEC)

MCSs shall:

(1) Coordinate interactions between multiple power sources
(2) Be listed, field labeled, or designed under engineering supervision
(3) Monitor and control power production and quality
(4) Monitor and control transitions with external primary power source

705 Part III Interconnected Systems Operating in Island Mode (New in 2023 NEC)

Here is some Article 705 history. The 2011 NEC 690.64 point of connection requirements were consolidated into 705.12(D) in the

2014 NEC. The 2014 NEC Article 705 was organized in three parts, Part I **General**, Part II **Utility-Interactive Inverters** and Part III **Generators**. In the 2017 NEC, 705 had the same three parts. In the 2020 NEC, Article 705 was reorganized into two parts, Part I **General** and Part II **Microgrid Systems**. Now in the 2023 NEC, we added another Part III again, in addition to the same parts we had in the 2020 NEC, which is short and called **Part III Interconnected Systems Operating in Island Mode**. This last new part simply restates the basics of Article 710, Stand-Alone Systems, to clarify that a multi-mode system that switches back and forth between grid-tied and islanded is NOT a stand-alone system anymore. When disconnected from the grid, it is an interconnected system operating in island mode. If you think this distinction is silly, or a waste of paper, you should sit in on the hours of discussion it took to get to that understanding—then you will really understand what wasted time looks like.

705.80 Power Source Capacity

When operating in island (stand-alone) mode the capacity shall be the sum of all of the power source maximum currents.

Discussion: It is often confusing when battery people talk about capacity, since Ah (ampere-hours) are considered capacity for batteries; however, in most cases capacity is expressed in power (watts). In this case it says we are expressing capacity in current as it is in Article 220.

Reminder: to get dc power we multiply current times voltage ($P_{dc} = I \times V$); to get ac power we multiply current times voltage times the power factor ($P_{ac} = I \times V \times p.f.$). An inverter is a volt-amp machine similar to an ac transformer. Proper rating of an inverter is in volt-amps rather than watts. If the power factor is low, the inverter cannot source any more volt-amps than when it is operating at unity power factor (p.f. = 1, no power factor loss). When we say we have a 7.6 kW inverter, what we really mean is that we have a 7.6 kVA inverter. Since grid-tied inverters operate at unity power factor most of the time, it is not inaccurate to say that the inverter is a 7.6 kW inverter most of the time. When the inverter is in grid-support mode and required to source or sink VARs, the inverter produces less power and no more than 7.6 kVA.

Alternating Confusion (For Non-Engineers)

kVA is called apparent power
VARs are called reactive power
Power is called real power
PF = power factor

With alternating current, current and voltage are often out of phase with each other and peaking at different times. Real power has to do with current multiplied by voltage instantaneously and is going to be equal to or less than apparent power. Apparent power multiplied by power factor equals real power.

Since ac nominal operating voltage is a given (either 120, 208, 240, 277, or 480Vac), then the actual capacity can be expressed by how much current the circuit actually draws. This is also helpful since it determines the size of the wire based on the made-up NEC word of "ampacity" which is the capacity of a wire to carry current at or below its voltage rating.

705.81 Voltage and Frequency Control

Voltage and frequency shall be supplied compatible with loads. This is an "apple pie and motherhood" statement that simply says that whatever power source we use, it must be compatible. The inverter and microgrid standards state exactly what is meant by compatibility. The NEC is all about safety. Compatibility is important so that equipment does not blow up, but the NEC is not the place to define compatibility, since that subject is complex. What we can say is that if a power source causes a healthy piece of electrical equipment to catch on fire, we should question whether the power source was truly compatible.

End-of-Chapter Thoughts

At this point, we have covered the meat and bones* of PV and the NEC with Articles 690 and 705. Rather than going **meat-and-bones**

level and creating a book that would be too heavy to carry and cover many things that are not specific to PV systems, we will go **broth level** on the rest of the Code.

*Solar planet-loving vegetarians: have no worries, the meat and bones used to make this book are actually soy-based, simulated meat products. Yum!

10 Storage Articles

As PV systems grow at an exponential pace and head towards the goal of someday saturating the grid, energy storage systems are a natural fit for bottling sunlight for nighttime photon usage (a.k.a. midnight solar).

This chapter will cover:

Article 706 Energy Storage Systems
Article 710 Stand-Alone Systems
Article 480 Stationary Standby Batteries
Article 625 Electric Vehicle Power Transfer System

With energy storage being in vogue at all the solar conferences this decade (or millennium), we will take a ride on the **Article 706 Energy Storage Systems** train and see where that track crosses paths with the **Article 710 Stand-Alone Systems** express at an intersection with the **Article 480 Stationary Standby Batteries** steamer.

Batteries and Energy Storage

In the 2014 NEC and earlier, **Article 480 Storage Batteries** was used along with **Section 690.10 Stand-Alone Systems** and **690 Part VIII Storage Batteries** for designing and installing energy storage for stand-alone PV systems. Much of that order was changed in the 2017 NEC. As usual, much of the change is with the organization of the NEC, rather than changing the installation drastically.

When work started on Article 706, the intention was to take the information from Articles 480 and 690 and get rid of those

DOI: 10.4324/9781003189862-11

sections. Somewhere along the way, the lead-acid folks decided they still needed Article 480 even though it is in the wrong chapter of the NEC (batteries do not belong in Chapter 4 in the first place) and 480 is no longer needed. The focus for field installations should be on Article 706 for energy storage systems that are used in and around utility services. The title and scope of Article 480 has changed to better represent the types of systems for which they are used. Article 480 is now called Stationary Standby Batteries as this equipment is mostly used by the telecom industry or for backup power in substations.

Scopes of Article 480 Stationary Standby Batteries and Article 706 Energy Storage Systems:

- Article 480 scope applies to all stationary standby installations of storage batteries.
- Article 706 scope applies to all permanently installed energy storage systems above 1 kWh.

As we can imagine, the scopes cross paths and both articles may be considered in battery systems with stored energy above 1 kWh.

Article 480, formerly known as Storage Batteries, has been in the NEC longer than we have been alive (Thomas Edison was in his prime) and Article 706 Energy Storage Systems is two Code-cycles old. As you read Article 480, you will realize that much of the material is related to good old time-tested and reliable (and sweet-tasting) lead-acid battery technologies. A new Informational Note in Article 480 seeks to differentiate between equipment for Article 480 and for Article 706.

Informational Note No. 1: See Article 706 for installations that do not meet the definition of stationary standby batteries.

To understand this note, we must look at the new definition in Article 100.

Definition: Battery, Stationary Standby. (Stationary Standby Battery)

A battery that spends the majority of the time on continuous float charge or in a high state of charge, in readiness for a discharge event. (CMP-13)

Informational Note: Uninterruptible Power Supply (UPS) batteries are an example that falls under this definition.

While it is likely that confusion will still occur between Articles 480 and 706, there is far more clarity in the 2023 NEC about which article to use for various applications. For purposes of this book, we will focus on applications covered by Article 706.

As we can see by reading Article 706, it focuses more on new modern technologies including, but not limited to, chemical batteries. Article 706 will cover the higher voltage lithium-derivative battery modules that are being sold, flywheel storage, capacitor energy storage, and compressed air storage. Article 480 will stick to standby battery applications that have been around for over 100 years.

Article 706 covers the energy storage "**system**." The International Residential Code (IRC) requires all energy storage systems used for one- and two-family dwellings to be listed to UL 9540. It is interesting to note that, at the time of this writing, there are still no UL 9540 listed lead-acid based ESS and that the IRC requires UL 9540. Some would say that this means you cannot install lead-acid batteries in a house anymore while others will claim that a house can have a stationary standby battery which is not considered an energy storage system. Clearly semantics will remain mirky on this issue. Ultimately, the local jurisdiction will have to decide whether or not to allow standby batteries in dwellings. Outside a dwelling is a completely different issue and is not specifically addressed in the IRC.

We will notice that there are many reoccurring themes in Article 706 from within the material in Articles 690 and 705. This includes having separate energy sources with the ability to operate in parallel and the requirements of disconnecting systems from each other.

An outline of noteworthy NEC **energy storage articles, parts, and sections follows**. *Italics are author comments.* We will start and emphasize Article 706 and end with Article 480, which is of less importance. The most important thing you can learn about Article 480 is how to explain that we generally do not use it. We will cover 706 first, since it is all important, and then cover 710 before we get to 480 and finish off with 625.

Article 706 Energy Storage Systems

706 Part I General

706.1 Scope

Article 706 applies to permanently installed energy storage systems over 1 kWh.

It is interesting that for the scope they also use the units of 3.6MJ, which equals 1 kWh. A joule is a watt second, so, since there are 3,600 seconds in an hour, 3.6 million joules = 1 kWh. You do not see as many people using joules in this industry as you do in physics classes. People with megajoules are rich and flashy dressers.

706.4 System Requirements (marking and labeling)

ESS shall be provided with a nameplate that is plainly visible and includes the following:

(1) ESS manufacturer's name or identifying information
(2) Frequency
(3) Number of phases (ac only of course)
(4) Rating in kW or kVA
(5) Available fault current at ESS output terminals
(6) Max output and input current at terminals
(7) Max output and input voltage at terminals
(8) Utility interactive capability

As we can see here, all the markings are not always required, since it is possible to have a dc energy storage system which would have no frequency or number of phases. It is also interesting to note that the power characteristics are required, but there are no energy characteristics required. It is nice to see that we do not have to calculate Ah! Now, when someone talks about Ah, you know they are trying to say that they have been around for a while and are probably a grumpy Article 480 diehard.

706.5 Listing

Energy storage systems must be listed!

This means UL 9540 and is not something you can make in your garage.

706.6 Multiple Systems

Multiple ESS can be on the same premises.

You may have to go to the residential, building, and fire codes to see different restrictions.

706.7 Commissioning and Maintenance

Since the NEC is not a maintenance code, we should not be looking in the NEC for maintenance requirements, we should be looking in the fire codes. Perhaps having these requirements in the NEC is a mistake since there is no way to enforce maintenance with a construction code.

706.7(A) Commissioning says that we need to commission ESS upon installation, but it does not apply to 1- and 2-family dwellings.

706.7(B) Maintenance tells us to keep our ESS maintained with written records and does not apply to 1- and 2-family dwellings.

706.9 Maximum Voltage

The maximum voltage is the rated input and output voltages on the nameplate or listing of the ESS. This is typically an ac voltage since it is coming out of the energy storage **system**.

706 Part II Disconnecting Means

706.15 Disconnecting Means

There are many similarities here when we look at 690.13 PV System Disconnecting Means.

706.15(A) ESS Disconnecting Means

Means to disconnect ESS from all wiring, power systems, and equipment shall be provided.

706.15(B) Location and Control

ESS shall be readily accessible and comply with **at least one** of the following:

(1) Located within ESS
(2) Located within sight and within 10 feet of ESS
(3) Lockable in the open position *(think lockout-tagout)*

Where remote controls of a disconnecting means are not within sight of the ESS, the location of the controls shall be marked on the disconnecting means.

Emergency Shutdown System: For 1- and 2-family dwellings, an initiation device readily accessible outside of the building shall plainly indicate whether on or off and when initiated (off) the ESS will stop exporting power.

Discussion: This is THE MOST IMPORTANT change in Article 706 in the 2023 NEC. **Emergency Shutdown System** sounds similar to 690.12 **Rapid Shutdown for PV Systems on Buildings.** It is possible to have the same "initiation device" control both the PV Rapid Shutdown and ESS Emergency Shutdown if it is marked for both. *This is good for firefighters!* This is also good for installers. The way the 2020 NEC is written, it is unclear whether the emergency disconnect, discussed in 706.15(B), is required to be a load-break switch or not. Without that detail in the 2020 NEC, and with the detail in the 2023, it is reasonable to make the case that the language of the 2023 NEC informs the language of the 2020 NEC as to what was intended. Having a shutdown control for an ESS is far easier to install than an external load-break switch.

706.15(C) Notification and Marking

ESS disconnecting means shall indicate whether on or off and be marked "ENERGY STORAGE SYSTEM DISCONNECT."

Disconnecting means shall also be marked to indicate:

(1) Nominal ESS voltage *(such as 240 Vac for your home)*
(2) Available fault current (*not much since typically coming from an inverter or dc-to-dc converter*)

(3) Arc-flash label (*more fault currents come from utility connection than ESS inverter*)
(4) Date calculation performed (*another throwback from Article 480*)

1- and 2-family dwellings can omit points (2) through (4), *so they only need nominal voltage.*

- Battery equipment suppliers can provide fault current information.
- NFPA 70E Standards for Electrical Safety in the Workplace has requirements for arc-flash labels.
    - *Unlike PV, which is current-limited, ESS can have excessive short-circuit currents which can be more dangerous with regards to arc-flashes; however, with electronics on the output of many UL 9540 listed ESS, we often do not see the currents like we did when we were connected to Article 480 style storage batteries.*

If the line and load terminals of the disconnect may be energized in the open position, then we need this familiar label:

WARNING
ELECTRIC SHOCK HAZARD
TERMINALS ON THE LINE AND LOAD
SIDES MAY BE ENERGIZED IN THE OPEN POSITION

706.15(D) Partitions Between Components

Where circuits from the terminals of an ESS go through a wall, floor, or ceiling, then a readily accessible disconnect shall be provided within sight of the ESS.

706.15(E) Disconnecting Means for Batteries (New in 2023 NEC)

If battery is subject to field servicing and separate from ESS electronics, then all of 706.15(E)(1) through (E)(4) shall apply.

Discussion: Many batteries that are separate from the inverter have ESS electronics and are not subject to these rules. Popular examples would be the LG Chem RESU and SolarEdge batteries.

706.15(E)(1) DISCONNECTING MEANS

Disconnecting means shall be provided for all ungrounded conductors, which is readily accessible and in sight of battery.

706.15(E)(2) Disconnection of Series Battery Circuits

Battery circuits shall be able to be broken into 240V dc or less circuits for maintenance. Non-load-break bolted or plug-in disconnects are acceptable.

706.15(E)(3) Remote Activation

If remote disconnect is out of sight from battery, then disconnect shall be capable of being locked in the open position.

706.15(E)(4) Notification

This is the same marking of nominal voltage, available fault current, arc-flash, and date of calculation requirements as 706.15(C) on pages 164 and 243, but without the 1- and 2-family dwelling exception.

706.16 Connection to Energy Sources

706.16 is disposable.

This whole section is unnecessary and duplicative since the requirements of Article 705 always apply when an ESS is connected to another energy source. The confusion probably comes from those ESS heretics who believe that ESS are not a source of energy themselves. These heretics believe that ESS are not a source—only a storage device. Regardless of their heretical beliefs, a charged battery looks like, quacks like, and in every way operates like other energy sources. Since 706.16 is referring to ESS connected to other sources, Article 705 clearly applies. Lastly, **706.16(F) Stand-Alone Operation** either is using the wrong language and reference or does not belong in this section. If the intent is to talk systems operating in island mode, that term should be used. If the intent is to talk about stand-alone mode, then it needs to be in a section other than 706.16. The best result is for the 2026 NEC to delete all of 706.16 except 706.16(F) and call the new 706.16 simply **Stand-Alone Operation**. Incidentally, if a battery is never connected to an energy source, we call that a disposable battery since it does not get recharged. It is clear that Article 706 is not intended to cover disposable batteries.

706.16 says to comply with all of 706.16(A) through (F) below

(A) **Source Disconnect**—If a disconnect has multiple sources, then it shall shut off all sources when in the off position.
(B) **Identified Interactive Equipment**—If we operate in parallel to ac sources, then equipment shall be listed and identified as interactive.
(C) **Loss of Interactive System Power**—If loss of primary power source, then interactive inverter circuit shall anti-island as described in 705.40 Loss of Primary Power Source (page 241).
(D) **Unbalanced Interconnections**—If unbalanced interconnection, then shall comply with **705.45 Unbalanced Interconnections** (page 232).
(E) **Other Energy Sources**—If connected to other energy sources, then shall be in accordance with 705.12 Interconnected Power Production Sources.
(F) **Stand-Alone Operation**—This section says that if the ESS can operate as stand-alone, then we follow **710.15 General (710 is Stand-Alone Systems)**. *However, this language is outdated and not consistent with the changes made in the 2023 NEC. If this section remains, then the heading should be "Island Operation" and state that if the ESS can operate in island mode, then follow **702.4(A) System Capacity**. Thanks for the additional confusion, Code Making Panel 13—they will figure it out eventually.*

706 Part III Installation Requirements

706.20 General (of ESS)

706.20(A) Ventilation

Ventilation shall be appropriate for battery technology (this repeats Article 480). Ventilation can be done according to manufacturer's instructions. *Lithium-ion battery systems require a battery management system (BMS) to prevent overcharging and they do not make hydrogen gas, like lead-acid batteries. Unless a lithium-ion ESS manufacturer states that ventilation must be provided, it is generally not required since sealed batteries generally do not require ventilation.*

706.20(B) Dwelling Units

An ESS shall not exceed 100Vdc between conductors or to ground, unless **during routine maintenance live parts are not accessible and then 600Vdc shall be acceptable**.

Many packaged ESS have dc voltages over 100V and have live parts that are not accessible during routine maintenance, which are hidden in the ESS. Lithium-ion batteries cannot be maintained so there is no reason to open a lithium ESS—unless you watched a YouTube video that taught you how to make your own Tesla Powerwall.

706.20(C) Spaces about ESS

706.20(C)(1) GENERAL

Repeats Article 480.10(C) in saying **Section 110.26 Spaces about Electrical Equipment** and **110.34 Work Space and Guarding**.

706.20(C)(2) SPACE BETWEEN COMPONENTS

Spacing is done in accordance with manufacturer's instructions. *After all it was tested and listed that way.*

706.21 Directory (Identification of Power Sources)

Labels shall be in accordance with 110.21(B) see page 269.

706.21(A) Facilities with Utility Services and ESS

Plaques in accordance with **705.10 Identification of Power Sources** (Interconnected Power Production Sources, pages 160, 169 and 194).

706.21(B) Facilities with Stand-Alone Systems

Plaques in accordance with **710.10 Identification of Power Sources** (Stand-Alone Systems, pages 160, 169 and 194).

706 Part IV Circuit Requirements

The circuit requirements are very much the same throughout the NEC, especially when calculating a wire size for a renewable energy system.

Here are a few things to take note of regarding current and an ESS.

- **706.30(A)(1) Nameplate-Rated Circuit Current**—Oftentimes an ESS will charge and discharge through the same conductors. If the charge and discharge rated currents are different, the greater current shall be used in wire sizing—*makes sense.*
- **706.30(A)(2) Inverter Output Circuit Current**—Continuous current.
- **706.30(A)(3) Inverter Input Circuit Current**—If the input of an inverter is connected to a battery, then the battery voltage will go up and down depending on the state of charge of the battery and the load put on the battery. We consider the **lowest voltage to determine the highest current**, since voltage × current = power and when voltage goes down, current goes up. We will discuss this further when we see it again soon on page 250 as we cover **Article 710 Stand-Alone Systems**.
- **706.30(A)(4) Inverter Utilization Output Circuit Current**—This is the stand-alone circuit and is based on continuous current (not surge current).
- **706.30(A)(5) DC to DC Converter Output Current**—Continuous current from dc-to-dc converter.

For more wire sizing, see Chapter 12 wire sizing, starting on page 287. The NEC repeats itself many times for wire sizing, and in this book we will be more practical and efficient.

- **706.31 Overcurrent Protection / 706.31(A) Circuits and Equipment**—Circuits shall be protected at the source of overcurrents. *Listed ESS have integral fault protection, so the dangerous fault currents are from the utility.*
- **706.33 Charge Control / 706.33(A) General**—We must have a means to control the charge and adjustable charge control is only accessible to qualified persons. *Listed ESS will have battery management systems to prevent overcharging.*
- **706.33(B) Diversion Charge Controller**—Diversion charge controllers are rare these days; however, if used there are requirements for oversizing diversion load and conductors. A diversion load is used to utilize otherwise wasted energy when the batteries are fully charged.
    - **706.33(B)(1) Sole Means of Regulating Charge**—We cannot rely solely upon a diversion charge controller to control a charge. If that diversion load were to break, then we would have nothing to prevent a perhaps dangerous overcharging of the battery.
    - **706.33(B)(2) Circuits with Diversion Charge Controller and Diversion Load**—We need to have:
        - Current of diversion load ≤ current rating of diversion load charge controller
        - Voltage rating of the diversion load > maximum ESS voltage
        - Power of diversion load must be ≥ 150% power of charging source
        - Ampacity diversion load circuit ≥ 150% max current diversion charge controller
        - OCPD diversion load circuit ≥ 150% max current diversion charge controller

If the diversion load is the utility, then we do not need to follow 706.33(B)(2) according to **706.33(B)(3) ESS Using Interactive Inverters** (too bad, I was hoping to use this to control the big bad utility company).

706 Part V Flow Batteries ESS

Flow batteries use pumps to move electrolyte and we do not see a lot of them, except in articles that claim there is a new technology that is going to take over, which rarely happens.

706.40 General (***Flow Batteries***)—Flow battery components must also follow parts **I General, II Disconnecting Means** and **III Installation Requirements** of Article 706. In the 2020 NEC we were, and no longer are directed to **Article 692 Fuel Cell Systems**. *A flow battery is like a fuel cell in many ways.*

706.41 Electrolyte Classification—Electrolyte shall be identified.

706.42 Electrolyte Containment—Electrolyte shall have containment to prevent spills.

706.43 Flow Controls—Controls to shut down provided in case of electrolyte blockage.

706.44 Pumps and Other Handling Equipment—Pumps rated for electrolyte.

706 Part VI Other Energy Storage Technologies

Part VI says to follow the Code and Parts I through IV of 706. Some other types of energy storage technologies that could fit here include flywheels, compressed air, capacitors, gravity storage, and whatever else some genius invents. Then 706 finishes with flywheels next.

706.51 Flywheel ESS (FESS)

FESS shall comply with the following.

(1) FESS not in 1- or 2-family dwellings *(that takes all the fun out of it!)*.
(2) FESS shall have bearing monitoring and controls.
(3) FESS shall have means to contain moving parts upon failure. This sounds like fun! An out-of-control flywheel can be like a bomb and cause serious injury. Most flywheels are buried.
(4) Spin down time shall be in maintenance documentation.

Article 710 Stand-Alone Systems

In the 2014 NEC, this material was covered in 690.10. Since then, the NEC has separated stand-alone systems from PV systems since there could be various power sources for a stand-alone system.

710.1 Scope

> This article covers electric power production systems that operate in island mode not connected to an electric utility or other electric power production and distribution network.
>
> Informational Note: These systems operate independently from an electric utility and include isolated microgrid systems. Stand-alone systems often include a single or a compatible interconnection of sources such as engine generators, solar PV, wind, ESS, or batteries.

This brief article covers electric power production sources that operate in island mode and not connected to a power production and distribution network.

Discussion: The 2023 NEC seeks to provide a better differentiation between stand-alone systems and other types of systems. Previously, a grid-tied battery backup system could be considered a stand-alone system when operating in island mode. This is no longer the case. With the changes to the Article 710 scope, and the new Part III of Article 705, Interconnected Systems Operating in Island Mode (see page 234), it is clear that Article 710 systems NEVER connect to a utility system. A remote stand-alone system that becomes interconnected with the utility grid moves from an Article 710 system to a Part III, Article 705 system.

Stand-alone mode *was officially changed to* ***island mode*** *in the 2020 NEC at the insistence of the islanders in Hawaii. They deserve it, given how often the power goes out there.*

Multimode inverters can operate in island mode or interactive mode, but if a multimode were never connected to the grid, then it would only operate in island mode.

710.6 Equipment Approval

Equipment must be listed, or field labeled. *Just because your ESS is off grid, does not mean you do not have to follow the rules. Don't use a non-UL listed truck stop 5kW 12V inverter for your homestead unless it is fireproof. This has been known to happen. That would be 417 Amps!*

710.10 Identification of Power Sources

A plaque, label, or directory will be at the power source's disconnecting means, or at an approved readily visible location. This will show each power source's disconnecting means location.

710.12 Stand-Alone Inverter Input Circuit Current

We mentioned (page 248) when we were discussing **706.30(A)(3) Inverter Input Circuit Current** that we would go into more detail here.

Maximum current shall be the continuous inverter input current when inverter is **producing rated power at lowest input voltage.**

Discussion: A stand-alone inverter takes power from a battery and the **battery will go through a range of voltages**, depending on whether the battery is **fully charged, charging** fast or slow, if the **load is surging** and the **minimum allowable voltage**, before the inverter will no longer take power from the battery in order to protect the battery from too deep of a discharge.

Since power = voltage × current, for consistent power, **as the battery voltage becomes lower, then the current will have to be higher in order to have consistent rated output power.**

For example, with a 1000W inverter, if the voltage is 14V as the battery is charging and we ignore inverter losses, then the required current would be calculated as:

1000W/14V = 71A

As the battery voltage goes down to 11.5A, then the calculation would be:

1000W/11.5V = 87A

If we had a 90% efficient inverter (hope it is better than 90%), then we could also calculate for the extra input current required to compensate for inverter losses to be a factor of 0.9.

87A/0.9 = 97A

Stand-alone inverters **require more input current at lower voltages to make the same amount of power**.

In the inverter installation manual, it may say that at a lower battery voltage, power (and thus current) will be reduced, so be sure to read the instructions. NEC 110.3(B) says we have to follow instructions, since that is how the equipment was tested and listed. Instructions, however, do not supersede the NEC; we have to follow both.

> **Stand-alone inverters** often have values for continuous current, which match the power rating of the inverter, but also have **current values for surges of current that are greater than the continuous current rating of the inverter**. For instance, it is not unusual for a 2kW stand-alone inverter to be able to handle a surge of 4kW for several seconds. This is because loads often require surge currents to get started. We size the wires based on the continuous currents and the stand-alone inverter output circuit maximum current, the wires and equipment are sized by the continuous current, which is less than the surge currents. **We do *not* size the wires based on the stand-alone inverter's greater surge currents**.
>
> Note that, in Figure 10.1 on the next page, there are power ratings that require much more current for time frames that are less than three hours (continuous). Wire sizes are based on currents that are continuous. 16.7A is based on 2000W at 120V. For short periods of time, there will be surges of double the current. Since the time is short, the wire will not have time to heat up and cause a problem.

Since we are on the subject, an inverter operating at full power will essentially have **no** surge capability, because even short surges will typically trip the inverter. To get the full surge capability, it is best to operate at half power. Using our example of a 2000W inverter, if we run it at 1000 watts, it will be ready to give us 4000 watts for short periods of time whenever we need it.

	Sealed
Models:	FXR2012A
Instantaneous Power (100ms)	4800VA
Surge Power (5 sec)	4500VA
Peak Power (30 min)	2500VA
Continuous Power Rating (@ 25°C)	2000VA
Nominal DC Input Voltage	12VDC
AC Output Voltage (selectable)	120VAC (100-130VAC)
AC Output Frequency (selectable)	60Hz (50Hz)
Continuous AC Output Current (@ 25°C)	16.7AAC
Idle Power	Full: ~34W Search: ~9W
Typical Efficiency	90%

Figure 10.1 Partial datasheet from outback stand-alone inverter.
Source: Courtesy Outback Power.

710.15 General

710.15 is the heart of Article 710 and Article 690 refers us to 710.15

710.15(A) Supply Output

Power supply can have less capacity than the calculated load.

Power supply shall be at least as much as the largest load. This is a bit of a *duh* statement, but it needs to be stated. Don't connect a 10kW load to a 5kW inverter. It will not work out for you. It's not unsafe unless you consider stupidity a hazard. Never forget that everyone is entitled to one fatal mistake. What will happen is the inverter will not be able to supply the load and will turn off. You may have to turn off the big load and push a reset button to get it working again.

Power supply can include multiple power supplies, such as solar, generator, and energy storage.

Oftentimes off-grid system owners will turn on a generator when using high power loads.

710.15(B) Sizing and Protection

Conductors between stand-alone source and disconnecting means shall be sized based on the sum of the output ratings of the stand-alone sources. Additionally, *and obviously*, 3-phase sources must be balanced to be compatible with loads.

When you are connected to a utility, your service conductors do not have to be sized based on the available current from the utility, but for a stand-alone source you have to size your conductors based on the source(s).

710.15(C) Single 120V Supply

Often a 120V inverter is used in stand-alone systems, and 120/240V designed service equipment **may be used** (bonding L1 to L2).

- The neutral bus must be rated greater than the sum of the loads.
- No multi-wire branch circuits since neutral currents will not cancel each other out and a sign must read:

WARNING:
SINGLE 120-VOLT SUPPLY. DO NOT CONNECT
MULTIWIRE BRANCH CIRCUITS!

710.15(D) Three-Phase Supply

Three-phase is allowed.

710.15(E) Energy Storage or Backup Power Requirements

"Energy storage or backup power supplies **not required.**"

Direct PV systems are stand-alone systems, which only work when the sun is shining.

710.15(F) Voltage and Frequency Control

Voltage and frequency shall be within limits compatible with loads.

Perhaps we can use a 230V 50Hz European inverter if we use compatible loads. This way we can have smaller conductors for the same amount of power. It is not all that practical as all receptacles would have to be special to prevent people from plugging 120V loads into 230V outlets. Perhaps your inner Francophile will love it though.

Article 480 Stationary Standby Batteries

480.1 Scope

This article applies to all installations of stationary standby batteries having a capacity greater than 3.6 MJ (1 kWh). *Remember that ESS are not stationary standby batteries whether you want them to be or not.*

> **Definition: Battery, Stationary Standby. (Stationary Standby Battery)**
> A battery that spends the majority of the time on continuous float charge or in a high state of charge, in readiness for a discharge event. (CMP-13)
>
> **Nominal Voltage (Battery or Cell)**
> Value assigned for convenient designation (nominal is in name only).
> Informational Note: Nominal voltages for different chemistries are listed as:
>
> Lead-acid 2V/cell
> Alkali 1.2V/cell
> Lithium-ion 3.2 to 3.8V/cell

It is a race between lead-acid and lithium that lead-acid has been winning since as early as 250 BC according to some sources. It looks as if that is changing with lithium battery technology going down in price and up in quality. As you can see, there is a range for the lithium-ion battery nominal voltage, which has to do with the cathode materials. Lithium iron phosphate (LFP) battery cells have a 3.2V nominal voltage, whereas lithium nickel manganese cobalt oxide (NMC) and lithium nickel cobalt aluminum oxide (NCA) cells have a 3.7V nominal voltage.

480.3 Equipment

Storage batteries and battery management equipment shall be listed, except for lead-acid batteries.

Interesting how we do not have to list lead-acid batteries.

480.6 Overcurrent Protection for Prime Movers

OCPD not required for 60V or less if battery provides power for starting, ignition, or control of prime movers.

*Discussion: A **prime mover** is the machine that turns the generator. An example of a prime mover is an internal combustion engine that spins something to make electricity. So, your starting battery circuit for your generator does not need overcurrent protection if under 60V. Other examples of prime movers are waterwheels, windmills, and steam turbines.*

480.7 Disconnect Methods

480.7(A) Disconnecting Means

Disconnecting means required for stationary standby battery ungrounded conductors with voltage over 60V.

*Apparently, battery systems below 60V are not required to have disconnects. This goes along with the overcurrent exception in 480.6. Energy storage systems however are required to have disconnecting means according to **706.15(A) ESS Disconnecting Means**, so it would be a good idea to be able to turn things off.*

480.7(B) Emergency Disconnect

1- and 2-family dwellings with a stationary standby battery system shall have a disconnect or remote-controlled disconnect outside of the building at a readily accessible location that is labeled "**EMERGENCY DISCONNECT**." This requirement is only relevant if an AHJ were to allow a stationary standby battery in a 1- and 2-family dwelling. These types of batteries would likely only be allowed in the attached garage of a dwelling. If the standby battery is located in a non-dwelling power shed for instance, no external disconnect is required because it is not a dwelling.

480.7(C) Disconnection of Series Battery Circuits

Battery circuits shall have disconnect provisions to break parts of the circuit into 240V dc segments or less. Non-load-break and bolted disconnects are acceptable.

480.7(D) Remote Actuation

If disconnecting means can operate with remote controls and controls are not within sight of the battery, then the disconnecting means shall be able to be locked on the open position. Location of controls to be marked on disconnecting means.

480.7(E) Busway

Disconnecting means can be incorporated into the busway.

480.7(F) Notification

Marking shall include:

(1) Nominal battery voltage (all that is required on 1- and 2-family dwellings)
(2) Available fault current (available from manufacturer)
(3) Arc-flash label
(4) Date short-circuit current calculation performed

Points (2), (3), and (4) are not required on 1- and 2-family dwellings. Only nominal battery voltage is required on 1- and 2-family dwellings.

480.7(G) Identification of Power Sources

This follows what we have already covered in different articles regarding plaques and directories. We are directed to 705.10 (pages 160, 169 and 194) and 710.10 (pages 169 and 252). There is an exception for cases where a 480.7(A) disconnect is not required (page 257).

480.9 Battery Support Systems

If using a corrosive electrolyte (like lead-acid), then support structure shall be resistant to corrosion. Metal structures shall have nonconducting support members or be constructed with insulating material. Paint alone is not sufficient.

480.10 Battery Locations

480.10(A) Ventilation

Ventilation appropriate to battery technology.

Lead-acid and nickel-cadmium batteries can create explosive hydrogen gasses.

Lithium batteries do not require ventilation as some inspectors once required since 480 used to call for it. Lithium batteries do not release hydrogen when overcharged, like lead-acid batteries do. Electrolysis is a way of making hydrogen with electricity. Water molecules are split into hydrogen and oxygen. ***If you stick a positive and a negative electrode into water, hydrogen will come off of the negative and oxygen will come off of the positive.***

480.10(B) Live Parts

Live parts shall be guarded in accordance with **Section 110.27 Guarding of Live Parts**.

480.10(C) Spaces About Stationary Standby Batteries

Spaces around batteries shall comply with **Section 110.26 Spaces About Electrical Equipment** and **110.34 Work Space and Guarding**.

For battery racks, there shall be a 1-inch space between a cell container and any wall or structure on a side not requiring access for maintenance.

This 1-inch space was on a national certification exam, so it must be important.

480.11(A) Vented Cells

Vented cells shall have flame arrestors (typical lead-acid).

480.11(B) Sealed Cells

Sealed cells shall have pressure release vents.

480.12 Battery Interconnections

Flexible cables shall be permitted and if used shall be used with proper lugs, terminals, etc. (rated for flexible cables) and cables shall be listed for the environmental conditions.

It has been a pet peeve for people writing exam questions to make sure that proper terminals are used for flexible fine stranded cables. This probably means a compression lug. If 480.12 did not exist, we would still have to do this, since it is obvious and included in other parts of the NEC. You do not have to use flexible fine stranded cable for batteries, but it can make maintenance easier, and it can help reduce mechanical stress on battery terminals.

480.13 Ground-Fault Detection

Battery circuits over 100V between conductors or to ground shall be permitted to work with ungrounded conductors if there is ground-fault detection and interruption.

This means that we are not required to have a grounded conductor.

Article 625 Electric Vehicle Power Transfer System is an article that will be used more and more by installers of PV and ESS equipment. Section 625.42 provided the basis for establishing 705.13 Power Control Systems in the 2020 NEC. In the 2023 NEC 625.42 was renamed Energy Management System (EMS) to match the changes to 705.13 (page 222) and 750.30 (page 223). Here is the language of 625.42 for reference:

> 625.42(A) Energy Management System (EMS)
>
> Where an EMS in accordance with 750.30 provides load management of EVSE, the maximum equipment load on a service and feeder shall be the maximum load permitted by the EMS. The EMS shall be permitted to be integral to one piece of equipment or integral to a listed system consisting of more than one piece of equipment. When one or more pieces of equipment are provided with an integral load management control, the system shall be marked to indicate this control is provided.

This language is critical to many residential systems that include PV and ESS since it allows the EMS to control the EV charger to keep the maximum load on a service or feeder within the ratings of the service or feeder. Without this language, it would not be clear that a 40-amp EV charger could ever be installed on a 100-amp residential service. **Few 100-amp services have the available capacity to add a level 2 EV charger without an EMS.** However, with an EMS to control the EV charger, the charger can be used if the EMS prevents overload of the 100-amp service. This means that if the load on the house is above 70-amps, the EV charge will either turn off or reduce power to the available capacity of the 100-amp service. This is a very big deal when adding EV chargers on 100-amp services.

Article 625 also contains section **625.48 Interactive Equipment**, which will direct **electric vehicles that can backfeed the grid** to Article 705. Section 625.48 also indicates that these **bidirectional batteries on wheels** need to be listed for exporting power.

When you are connecting your electric vehicle to your house as an optional standby system, you are encouraged to follow the Code and use **Article 702 Optional Standby Systems**. Article 702 is most often used when backing up a building or a house with a generator. There are some YouTube videos that show people how to connect electric cars to their houses to operate in stand-by mode in a non-Code-compliant way. Be careful! Do not have too much fun! Article 702 has historically been about connecting a backup power source through a transfer switch. The 2023 NEC changed all that. Now Article 702 is supposed to be used for transfer switches **and microgrid interconnect devices (MIDs)**. Most PV systems used for backup power connect to a multimode inverter or work with an MID. The requirements for automatic load connections in 702.4(B)(2) require that we use an energy management system (EMS) to control the loads connected to the microgrid. Fortunately, **many microgrid systems are being listed as PCS or EMS which make them compliant with 702.4**.

As you and your customers contemplate implementing energy storage into a grid-connected PV project, be sure to understand the regulations and incentives for connecting energy storage to the grid. Policies are evolving, and the Public Utilities Commissions are working hard to determine how energy storage will be brought into the grid. A good source for finding out about incentives for all things renewable is the Database of State Incentives for Renewables and Efficiency at www.dsireusa.org, which is from North Carolina, just like Bill. In fact, he helped start the organization that developed and maintains this database.

11 Chapters 1–4, Chapter 9 Tables, and Informative Annex C

So far, we have covered the NEC articles that are used often in the solar industry, including 690, 705, 480, 706, 710, and 625. These articles often refer to other articles in the NEC and this chapter will go over the different articles of the NEC from a PV perspective.

First, we will cover Chapters 1–4 of the NEC, which apply generally to all electrical installations. We will then mention the relevance of, and when to use, the articles in Chapters 5–7. Finally, we will cover Chapter 9 and look at the informative annexes that we can use when designing PV systems.

Chapters 1–4 have been referenced throughout this book, especially with respect to Articles 690 and 705.

Here are some of the top articles to be familiar with in Chapters 1–4:

- Article 100 Definitions (huge in 2023 NEC)
- Article 240 Overcurrent Protection
- Article 250 Grounding and Bonding
- Article 310 Conductors for General Wiring [especially the 310 tables!]
- Article 358 Electrical Metallic Tubing: Type EMT

Chapters 1–4 of the NEC

The NEC is divided into chapters, and the number of the chapter precedes the three-digit number of the article. For instance, Chapter 6 Special Occupancies includes Article 690 Solar PV Photovoltaic Systems and Chapter 1 General begins with Article

DOI: 10.4324/9781003189862-12

100 Definitions. Chapters 1–4 apply to all electrical installations, including PV systems.

An outline of **often-used-for-PV articles, parts, and sections** in Chapters 1–4 of the NEC follows. *Italics are author comments.*

Recall the hierarchy of NEC organization: Chapters/Articles/Parts/Sections.

Chapter 1 General

Article 100 Definitions

Article 100 is now the longest article in the NEC, since in the 2023 NEC all of the ".2" definitions were moved to Article 100, such as 690.2, 705.2, 706.2 etc. Rather than cutting and pasting all the definitions in Article 100 and making this book twice as heavy, we recommend going straight to the source. The definitions hopefully are self-explanatory and if not, we will cover some of the confusing definitions here, to make them readily accessible to your brain.

Article 100 is a great resource for defining terms throughout the NEC, including "**PV**."

*First off, in Article 100 definitions, we see **three types of accessible**, one being **applied to equipment**, another **applied to wiring methods**, and the third is the well-used term **readily accessible**. Let's start off our definitions with the three "accessibles," just as Article 100 starts off with these "accessibles."*

Accessible (as Applied to Equipment)

"Capable of being reached for operation, renewal or inspection."

Remember that accessible equipment can be reached.

Accessible (as Applied to Wiring Methods)

"Capable of being removed or exposed without damaging the building structure or finish or not permanently closed in or blocked by the structure, or other electrical equipment, other building systems, or finish of the building."

Remember that if it is a wiring method and you have to dismantle something, then it is not an accessible wiring method.

Accessible, Readily

"Capable of being reached quickly for operation, renewal, or inspections without requiring those to whom ready access is requisite to take action such as to use tools (other than keys), to climb over or under, to remove obstacles, or to resort to portable ladders, and so forth."

*Remember that **readily accessible** means you can get at something with a key or less and do not need tools or a ladder to get at something.*

Branch Circuit, Multiwire (Multiwire Branch Circuit)

A branch circuit that consists of two or more ungrounded conductors that have a voltage between them, and a neutral conductor that has equal voltage between it and each ungrounded conductor of the circuit and that is connected to the neutral conductor of the system.

This is referenced in Article 710 Stand-Alone Systems: When you have a 120V system and are using typical 240V split-phase equipment, you are not allowed to use multiwire branch circuits, since you may overload the neutral.

Converter, Dc-to-Dc (Dc-to-Dc Converter)

A device that can provide an output dc voltage and current at a higher or lower value than the input dc voltage and current.

*Discussion: The dc-to-dc converter ("**optimizer is a SolarEdge term**") definition was put into the Code in 2014.*

Power Optimizers

Dc-to-dc converters are commonly referred to in the industry as "power optimizers," which is really a marketing term from SolarEdge. If a dc-to-dc converter did not work as well as advertised or was clipping power (reducing power on purpose), perhaps we would call it a "power-pessimizer."

We could also have dc-to-dc converters that are not associated with PV and were part of an energy storage system.

In a way a maximum power point tracking (MPPT) charge controller is a dc-to-dc converter taking one voltage in and another voltage out. Also, within our string inverters electronically, there is usually one or more dc-to-dc converters and we could call each separate MPPT in the inverter a dc-to-dc converter. A transformer converts ac voltage and current to a different level of ac voltage and current, like a dc-to-dc converter converts dc voltage and current to a different level. Perhaps in the future we can call a transformer an ac-to-ac converter and an inverter a dc-to-ac converter.

Electronic Power Converter

A device that uses power electronics to convert one form of electrical power into another form of electrical power.

Discussion: An inverter or a dc-to-dc converter is an EPC; however, a transformer is not, since it uses magnetics instead of electronics.

Energy Management System (EMS)

A system consisting of any of the following: a monitor(s), communications equipment, a controller(s), a timer(s), or other device(s) that monitors and/or controls an electrical load or a power production or storage source.

Discussion: This is a new definition and as we discussed on page 222 the 2023 NEC has changed 705.13 Power Control Systems to 705.13 ***Energy Management Systems (EMS)****. EMS was already Article 750 in the NEC; however, now, as you can see from this new definition,* ***an EMS can control sources****, in addition to loads.*

Energy Storage System (ESS)

One or more devices installed as a system capable of storing energy and providing electrical energy into the premises wiring system or an electric power production and distribution network.

Discussion: In an Informational Note, it explains that the ESS differs from a stationary standby battery installation, since the later spends most of its time in float charge or a high charge in readiness for a discharge event. This Informational Note is an interesting distinction; however, many ESS are used as a standby source and not cycled unless there is an outage. The Tesla Powerwall at my mom's house is never cycled down unless the grid goes down and I would still call it an ESS, so perhaps this Informational Note is not entirely accurate.

Generating Capacity, Inverter

This is the output of the inverter. This is measured in kW, W, kVA or VA and at 40°C. We often call this our ac system size. **690.7 Maximum Voltage** and **690.8 Circuit Sizing and Current** have some special exceptions for systems with a generating capacity over 100kW and **Article 691 Large-Scale Photovoltaic (PV) Electric Supply Stations** has exceptions that apply to systems with a generating capacity of 5000kW (5MW) or greater, among other criteria.

Inverter, Interactive (Interactive Inverter)

Discussion: On the street they call this a grid-tied inverter. **Interactive inverters** operate in what can be called **current-control** mode (a.k.a. grid-following mode). A **stand-alone inverter** operates in **voltage-control** mode (a.k.a. grid-forming mode).

Inverter, Multimode

A multimode inverter can work in **interactive** mode (**current-control** mode), or it can work in **island** mode (**voltage-control** mode). A multimode inverter will typically have **different outputs for the interactive and island-mode circuits**. A multimode inverter has also been called a bimodal inverter in some books since there are normally two primary modes (current mode and voltage mode). Very often multimode inverters are **incorrectly (to the NEC) called hybrid inverters**. Inverters may not be hybrid. A NEC-defined hybrid PV system will have another source of power **besides PV or batteries,** such as a generator.

Island Mode

The operating mode for power production equipment or microgrids that allows energy to be supplied to loads that are disconnected from an electric power production and distribution network or other primary power source. (CMP-4)

Microgrid

An electric power system capable of operating in island mode and capable of being interconnected to an electric power production and distribution network or other primary source while operating in interactive mode, which includes the ability to disconnect from and reconnect to a primary source and operate in island mode. (CMP-4)

Discussion: This definition has always seemed as if you could not have a microgrid, without a grid to connect and disconnect from; however, the first Informational Note in the scope of Article 710 shows that isolated microgrids are stand-alone systems.

"Informational Note: These systems operate independently from an electric utility and include isolated microgrid systems. Stand-alone systems often include a single or a compatible interconnection of sources such as engine generators, solar PV, wind, ESS, or batteries."

Microgrid Interconnect Device (MID)

A device that enables a microgrid system to separate from and reconnect to an interconnected primary power source.

Power Production Equipment

Electrical generating equipment supplied by any source other than utility service, up to the source disconnecting means.

PV (Photovoltaic) System (PV System) (Photovoltaic System)

"The total components, circuits, and equipment up to and including the PV system disconnecting means that, in combination, convert solar energy into electric energy."

The PV system used to include the energy storage system and even some loads in the 2014 NEC and earlier. Since renewable energy is taking over, we get to have more specialized code and separate articles for these different systems. We can often say that the separation point between the PV system and something else is the disconnecting means closest to the point of interconnection for an interactive system. For a dc coupled stand-alone system, this disconnect would be the dc disconnect going to the energy storage system. A PV system does not include storage or loads.

Service

"The conductors and equipment connecting the serving utility to the wiring system of the premises served." *Electricity is served!*

Service Conductors

"The conductors from the service point to the service disconnecting means."

On your house this would be between the meter (service point) and the main breaker. We typically make the source connection to a service (supply-side connection) on the service conductors.

Service Drop and Service Lateral

"The conductors between the utility supply system and the service point."

A service drop is overhead, *and a* ***service lateral is underground.*** *Often coming from a transformer to your meter.*

Service Equipment

Main disconnect and associated equipment.

Stand-Alone System

Synonym: Off-grid.

NEC wording: "A system that is not connected to an electric power production and distribution network."

Note: Article 710 is Stand-Alone Systems.

Article 110 Requirements for General Installations

Article 110 is mostly used for electrical connection rules, working spaces around equipment, and enclosure types.

110.14(C)(1) covers the terminal temperature ratings as they relate to equipment ratings and conductor ratings. It tells us that we can only operate a conductor up to its current ampacity (at

30°C) for the temperature of the terminal to which it is connected. Therefore, a 10 AWG THWN-2 conductor can be operated at no greater than 35A if the wire is connected to a 75°C terminal on a circuit breaker. (Stay tuned for wire sizing examples in next chapter.)

110.21(B) Field-Applied Hazard Markings

Often referred to in 690 and 705. Labels meeting these requirements shall:

(1) Warn of hazard using effective words, colors or symbols
(2) Be permanently attached and not handwritten unless handwritten information is subject to change

There are ANSI (American National Standards Institute) standards **recommended in Informational Notes** that go into colors, font sizes, wording, design, and durability.

110.26 Spaces about Electrical Equipment

This is often **3 feet back and 30 inches wide** or the width of the equipment for wider equipment. Also **6.5 feet high** or height of equipment, whichever is greater. Table 110.26(A) has different spaces for higher voltage equipment and if there are exposed live parts, the distances can be greater. Recall that exposed live parts means: "Capable of being inadvertently touched nearer than a safe distance by a person."

Table 110.28 Enclosure Types

This includes NEMA enclosure types for wet areas, wind, dust, etc. Remember that any NEMA with a 3 or 4 in it is most commonly used outdoors, along with other NEMA ratings.

Chapter 2 Wiring and Protection

Article 200 Use and Identification of Grounded Conductors

200.6 Means of Identifying Grounded Conductors

200.6 is where we are told to have a usually **white** or sometimes gray (or other) **grounded conductor**. [See 690.41(A)(5) page 134].

A grounded conductor is a conductor that can carry current but is "solidly" connected to ground and operates at the same voltage as ground. Applying 200.6 on a dc PV circuit is way less common for 2023 NEC (2017 NEC and after) compliant PV systems than for 2014 NEC and earlier compliant PV systems since the functionally grounded array definition requires no dc grounded conductor marking. See page 128. We still do have white (or gray) colored grounded conductors on the ac side when inverters require a neutral, just not on the dc side.

Article 230 Services

Source connections to a service (supply-side connections) need to be properly bonded using best practices like service equipment since they are exposed to utility currents. In the 2023 NEC, these provisions are found (hidden) in **250.25 Grounding of Systems Permitted to be Connected on the Supply Side of the Service Disconnect**. We have referred to and explained some of the information in Article 230 when covering 705.11 (page 197). Services typically have no OCPD coming from the utility and have enough short-circuit current to vaporize metal in some cases. Vaporizing metal is not recommended. (It is the worst kind of vaping.)

In earlier versions of the NEC, there was confusion about methods of bonding the neutral of a supply-side connection. In the 2023 NEC, it is stated that we are to use Parts II through V and VIII of Article 250. Solar is not a service, but a **source connection to a service** is treated like a service when it comes to the point of system grounding.

Article 240 Overcurrent Protection

240.4 Protection of Conductors

240.4(B) OVERCURRENT DEVICES RATED 800 AMPERES OR LESS

You can protect a conductor with an OCPD that is the next higher standard overcurrent device size above the conductor ampacity. This is the part of the following chapter on wire sizing that can be confusing because it allows a conductor to have an ampacity less than the OCPD rating that is protecting it. Sometimes installers think that the OCPD has to be less than the short-circuit current or

125% of continuous current, which is not the case, since we do not want false trips. 240.4(C) tells us that, once your OCPD is 800A or more, the ampacity of the conductor must be no smaller than the OCPD rating. You can never have too much ampacity, unless you are trying to save money or space.

240.4(D) SMALL CONDUCTORS

"*Small conductor rule*":

- 14 AWG copper wire needs at least 15A overcurrent protection.
- 12 AWG copper wire needs at least 20A overcurrent protection.
- 10 AWG copper wire needs at least 30A overcurrent protection.

There is more information in 240.4(D) about small aluminum conductors in the NEC.

Many indoor electricians go by the small conductor rule for wire sizing, since they do not have high ambient temperatures and are just running NMC (Romex) to breakers all day long. Many solar electricians use larger conductors because of voltage drop (efficiency).

240.6 Standard Ampere Ratings

- Special fuses
- 1A, 3A, 6A, and 601A (not used for us much)
- Fuses and breakers standard sizes from Table 240.6(A)
- 5A increments 10 to 50A
- 10A increments 50 to 110A
- 25A increments 125 to 250A
- 50A increments 250 to 500A

[See Table 240.6(A) for larger OCPD sizes up to 6000A.]

We are allowed to use a larger OCPD than calculated if the installation instructions of the equipment, such as the inverter, say that we can (if it was tested that way). For instance, since 25A breakers are difficult to find, we can often use a 30A breaker as long as the wire sizing checks work, as we will see in the next chapter. There are now larger 100A–300A inverters that can use a 1000A fuse in its listing and instructions, which allows multiple inverters on a single set of fuses (much like we already do for microinverters).

240.21(B) FEEDER TAPS

We covered this when we covered 705.12(B)(2), pages 161, 214 and 221.

- 240.21(B)(1) 10-foot tap rule (we used the 10-foot tap rule for interactive inverters in the 2014 to 2020 NEC, but **not in the 2023 NEC**, where we now use the 1/3rd calculation from the 25-foot tap rule below). The 10-foot tap rule was not often a factor anyway.
- 240.21(B)(2) 25-foot tap rule (for conductors 25 feet or less, we can use 1/3rd of the ampacity in our calculations [see 705.12(A)(2) on page 208].
- 240.21(B)(4) Taps in high bay manufacturing buildings use the 1/3rd calculation and can go up to 25 feet horizontally and 100 feet in total length.

Article 250 Grounding and Bonding

690 Part V is also Grounding and Bonding; see page 126. **690 Part V** often refers to 250.

250 Part III Grounding Electrode System and Grounding Electrode Conductor

This is where experts differ and there is confusion. You could say that with different conditions, such as lightning and the variable resistance of earth, there is no perfect solution to grounding, which is why there may be so many different opinions. **Lightning protection is not covered in the NEC** and is covered in **NFPA 780: Standard for the Installation of Lightning Protection Systems**, which is not something that AHJs typically adopt. Different AHJs will have different ways of doing things, especially when it comes to grounding.

250.52(A) Grounding Electrodes Permitted

(1) metal underground water pipe (*common*)

(2) metal in-ground support structure
(3) concrete-encased electrode (*common*)
(4) ground ring
(5) rod and pipe electrodes (*common*)
(6) other listed electrodes
(7) plate electrodes (*common in Canada*)
(8) other local metal underground systems (underground metal well casing, tanks, piping systems, etc.)

250.52(B) Not Permitted for Use as Grounding Electrodes

(1) metal underground gas piping systems
(2) aluminum
(3) structures and structural rebar in 680 (Swimming Pools)

250.53 Grounding Electrode System Installation

Learn how to install electrodes here and perhaps memorize where to go.

Table 250.66 Size of AC Grounding Electrode Conductors (GEC)

The size of the ac grounding electrode conductor (GEC) is based on the size of the largest ungrounded service entrance conductor.

Usually, when dealing with PV on an existing service, you already have an existing ac grounding electrode conductor and do not need to go to **Table 250.66**.

Part VI Equipment Grounding and Equipment Grounding Conductors

250.122 Size of Equipment Grounding Conductors (EGC)

EGC sizes are based on the size of the overcurrent protection device. See page 146. This table address of **250.122** may be worth memorizing for some since it is commonly used for ac and dc. It is interesting to some that the EGC can be much smaller than the current-carrying conductor for PV systems, and also for PV systems we are not required to upsize the EGC when we upsize the EGC for voltage drop reasons, as we discussed on page 145 where we covered **690.45 Size of Equipment Grounding Conductors**.

250.166 Size of the Direct-Current Grounding Electrode Conductor (GEC)

For most PV systems, including functional grounded systems, there are no dc grounding electrode conductor requirements. In these systems, the **ac equipment grounding conductor and ac grounding electrode conductor provide reference and a pathway to the grounding electrode system.**

690.47(A)(1) "**For PV systems that are not solidly grounded, the equipment grounding conductor for the output of the PV system, where connected to associated distribution equipment connected to a grounding electrode system, shall be permitted to be the only connection to ground for the system.**"

In past Code cycles, for you old-timers, it was common for an AHJ to require a separate dc electrode or dc grounding electrode conductor for most PV systems. It is now silicon crystal clear that this is no longer the case.

An example of a "solidly grounded" PV system that requires a dc grounding electrode conductor is a typical dc direct PV water pumping system.

Chapter 3 Wiring Methods and Materials

Article 300 General Requirements for Wiring Methods and Materials

300.7 Raceways Exposed to Different Temperatures

300.7(A) SEALING

If a raceway is going to areas of different **temperature changes**, seal the raceway at the junction of the temperature change to **prevent condensation** where warm air would otherwise meet cold pipe. *In the 2020 NEC it was added, and kept in the 2023 NEC, that the sealant shall be identified for use with cable insulation, conductor insulation, a bare conductor, or other components. Some molecules do not play well with others. Duct seal is often used.*

300.7(B) EXPANSION

If a raceway is long enough and exposed to enough changes in temperature, calculations like voltage temperature calculations can be made to determine the need for expansion joints or flexible conduit. These calculations also apply to the expansion of solar rails.

We can take the difference between the low and high temperature and then multiply the difference in temperature by a coefficient to get the expansion.

A PV fire was once blamed by some on expansion joints scraping the insulation off a wire on a PV output circuit. Some electricians will use some flexible conduit between stretches of rigid conduit to allow for expansion, rather than expansion joints.

Article 310 Conductors for General Wiring

This is where we do a lot of our wire sizing, which we will do in the next chapter. 310 is a very popular place in the NEC.

310.14(A)(2) Selection of Ampacity (plus exception)

If different sections of a circuit have different ampacities, then the lowest ampacity shall apply unless the shorter section is **less than or equal to 10 feet or 10% of the circuit length, whichever is less.**

A conductor can be like a heat sink and the shorter piece can share its heat with the longer cooler piece. (Just as a conductor can share its coolness with a terminal.) 😎

310 Ampacity Tables (includes the famous Tables 310.16 and 310.17)

310 Tables [most often used pages in the NEC!]. For examples see Chapter 12 "Wire Sizing" of this book.

Table 310.15(B)(1)(1) Ambient Temperature Correction Factors Based on 30°C (formerly Table 310.15(B)(2)(a) in 2017 and 310.15(B)(1) in 2020)

Used to derate conductor ampacity in hot places, such as outside in the summer. (Someone in charge of 310 is making themselves feel important by changing table names every cycle.)

It is recommended by the authors to use **ASHRAE 2% average high temperature data** found at www.solarabcs.org/permitting. The more conservative 0.4% average high temperature data is not required.

Table **690.31(A)(3)(2) Correction Factors** (page 103) is like an extension of 310.15(B)(1)(1) and shall be used for solar PV system wires rated for 105°C and 125°C, just in case you are planning on running your PV circuits through a sauna, or want to use a really small conductor.

Table 310.15(C)(1) Adjustment Factors for More than Three Current-Carrying Conductors

For four or more conductors in conduit or cable, since they will have trouble dissipating heat and shall be derated. An analogy here is a real Alaskan sleeping with four dogs on a cold night to stay warm, which is also called a four-dog-night.

Not to be applied if distance is less than 24 inches according to **310.15(C)(1)(b).**

Not to be applied to neutrals carrying only unbalanced currents according to **310.15(E)(1).**

Removed Table 310.15(B)(3)(c) Ambient Temperature Adjustment for Raceways or Cables in Sunlight on a Roof (last seen in the 2014 NEC and has been *removed*).

310.15(B)(2) Rooftop (not a table) is in the 2023 NEC and says that if our raceways or cables are **3/4 inch or less above rooftop,** we shall add a 33°C adder. It was 7/8 inch in the 2020 NEC. It is **recommended** to be at least 1 inch above rooftop to avoid the adder and **to keep debris from building up on the rooftop.**

For PV circuits in cable trays, refer to **690.31(C)(2) Cable Tray** on pages 110–114.

Table 310.16 Allowable Ampacities of Insulated Conductors [not in Free Air]

For conductors, such as in raceway, cable, or buried.

This is the most used page in the NEC. It gives the ampacities of conductors in cables, raceways, direct buried, and everything else that is not in free air.

Table 310.17 Allowable Ampacities of Insulated Conductors

For conductors in free air (like the PV wire under an array).

310 TABLES DISCUSSION

You will get to see us using all these tables in the next chapter on wire sizing. Tables 310.16 and 310.17 have columns that have temperature ratings of conductors on them. Some of the commonly used conductor insulation types are listed, such as THWN-2 and USE-2, which are commonly used wires for PV systems and 90°C rated. This means that they can have the amount of current going through them in the column for the specific wire size without having a chance of going over the temperature rating. We are not allowed to go over the temperature rating of the insulation of the conductor. Once again, it is best to see our examples in the next chapter, Chapter 12 Wire Sizing, and then follow them to size your own wires and OCPDs using the same pattern, hopefully all in the same sitting, until you get the hang of it. Repetition helps here, especially when coupled with **common sense**, which is almost as rare as a **105°C or 125°C rated conductor**, which you can find an **ampacity table for in 690.31(A)(3)(1)** covered on page 101. Recall that 690 is only for PV systems.

In Table 310.16, look at a 10 AWG THWN-2 and you will see in the 90°C insulation column that conductor would have an ampacity of 40A when at 30°C (the entire table is for 30°C). If it is hotter than 30°C or if there are more than three current-carrying conductors in a raceway, then we would have to correct the 30°C value to the lower value and have a lower conductor ampacity.

Articles 320–362 Various Cables and Conduits

From EMT to NMC (Romex), you will find it here.

Article 330 Metal-Clad Cable: Type MC

MC Cable is a cable assembly sold with conductors inside of an armor of interlocking metal tape. MC Cable can be installed in a wet location if the conductors under the metallic covering are listed for wet.

MC Cable or raceway is cited in **Article 690.31(A) Wiring Systems** for an option for a wiring method for PV source and output circuits in readily accessible locations. Also see page 100.

MC Cable or metal raceway is a wiring method for PV dc circuits inside of buildings that is cited in **690.31(D) Direct-Current Circuits on or in Buildings**. Also see page 118.

MC Cable is in no way related to MC connectors. MC (Multi-Contact) connectors are made by the company Stäubli, a Swiss company that has not quite a monopoly on the PV connector industry according to MC Hammer (a rapper born in Oakland, just like Sean).

Article 334 Non-Metallic-Sheathed Cable: Types NM and NMC [Romex]

Also commonly known as **Romex**, which is a brand name of wiring commonly seen in residential buildings. You can have your ac inverter circuits in Romex where NM and NMC can be used.

This is a factory-assembled cable with insulated conductors within a non-metallic jacket. You will see a lot of this going in and out between the studs of a house being built. It is usually yellow or white.

Romex cannot be installed in storage battery rooms.

Article 336 Power and Control Tray Cable: Type TC

Factory-assembled of two or more insulated conductors under a non-metallic jacket.

Article 338 Service-Entrance Cable: Types SE and USE

This is where we learn about the **USE-2** that we use in our dc circuits under the array which also **must be listed as RHW-2** at the same time, which is normal for most USE-2 wire. We can see more about USE-2 uses in PV systems in **690.31(C)(1) Single-Conductor Cable** (see page 108).

Article 338 says that USE cable cannot be used for interior wiring; however, when it is dual listed as RHW-2, then it can be used indoors, so it really can be used indoors, since it is almost always dual listed as RHW-2. The requirement for USE-2 to be dual listed as RHW-2 is new in the 2020 NEC. It is funny how this wire that starts with U for underground is sunlight resistant. *Perhaps it is for those neutrinos that can penetrate a planet.*

Article 342 Intermediate Metal Conduit: Type IMC

IMC is a steel threadable raceway which is good at providing physical protection and for the most part can be used interchangeably with RMC (next).

In some AHJs, such as many in the Chicago area, EMT is not allowed for solar circuits and everyone is using more heavy-duty IMC. An AHJ can make up their own rules and not follow the NEC whenever they want. AHJs are all powerful, which is one reason why the cost of bureaucracy can be $1/watt in some instances according to SEIA (www.seia.org/initiatives/solar-automated-permit-processing-solarapp). Praise be the AHJ!

Article 344 Rigid Metal Conduit: Type RMC

RMC can be used in the same manner as IMC (above). IMC is easier to bend and work with and RMC is thicker.

Article 350 Liquidtight Flexible Metal Conduit: Type LFMC

Liquidtight Flexible Metal Conduit is a metallic raceway, which can be used for dc PV circuits inside and outside of buildings. Some solar installers will use the more expensive, yet easier to bend LFMC rather than EMT. Note that most LFMC is rated 60°C when wet. LFMC shall be supported every 4½ feet.

Article 352 Rigid Polyvinyl Chloride Conduit: Type PVC

PVC is most often used in Hawaii outside of buildings due to high corrosion rates for metal. PVC cannot be used inside of buildings for dc PV circuits, since it is not metal. It is easily broken in cold weather. *Table 352.30(B) Support of Rigid PVC tells us how often to support different sizes of PVC. ¾ inch PVC is supported every 3 feet for example.*

Article 356 Liquidtight Flexible Non-Metallic Conduit: Type LFNC

Liquidtight Flexible Non-Metallic Conduit looks a lot like Liquidtight Flexible Metallic Conduit from the outside, since the metal version, which we just covered in Article 350, has the metal inside of plastic.

Many times, you will see solar installers connecting inverters to junction boxes, disconnects, monitoring and other equipment using LFNC on the outside of a building. You see electricians in the field refer to this wiring method as just **liquidtight**. It is not for PV dc circuits inside of a building, since it is not metal.

Article 358 Electrical Metallic Tubing: Type EMT

EMT conduit is the most often used wiring method among solar installers on the mainland United States. EMT takes some practice to be good at bending. Journeyman electricians have EMT bending competitions and champions. You can get 10-foot sticks of ¾ EMT for just a few bucks (depending on supply-chain issues).

EMT is typically a thin-walled steel raceway that we will fill with conductors, such as THWN-2.

EMT can be inside or outside and where exposed to physical damage.

EMT is not threaded and we will use watertight connectors when outside (wet location).

Although we may be using EMT that is in a wet location with watertight connectors, we still need to have a conductor inside the EMT that is listed for wet locations. When conduit will go outside the conductors inside it will have a good chance of getting wet be it from rain or condensation.

358(A) Securely Fastened

The EMT is to be secured every 10 feet (3 feet from junction boxes and equipment).

Article 392 Cable Trays

Cable trays are often used around arrays to manage wires. PV wire is allowed in cable trays, as we can see in **690.31(C)(2) Cable Tray** on pages 110–114.

Chapter 4 Equipment for General Use

Article 400 Flexible Cords and Cables

Referenced earlier in this book on pages 115 and 116, these cables are often used for solar trackers, which follow the sun, and for storage batteries. See **690.31(C)(4) Flexible Cords and Cables Connected to Tracking PV Arrays.** Additionally, see Table **690.31(C)(4) Minimum PV Wire Strands** and **690.31(C)(5) Flexible, Fine-Stranded Cables.**

Article 480 Storage Batteries

See Chapter 10 storage with Article 480 Stationary Standby Batteries starting on page 256.

Article 495 Equipment over 1000 V ac, 1500 V dc, Nominal [formerly Article 490 in 2020 NEC]

490.2 High voltage is defined as 1000V ac and 1500V dc for this article. In the previous version of the NEC, it was just 1000V (did not specify ac or dc; however, 690 made it 1500V dc and modified Chapter 4).

For PV systems over 1000V, the **2023 NEC 690.31(G) Over 1000V** was covered on page 122.

Complying with Article 495 is possible, yet difficult, which is why we are not seeing 2000V PV arrays yet, at the time of this writing, but stay tuned…

Chapter 5 Special Occupancies

If you are installing a PV system in a barn, on a gas station, an aircraft hangar, or other special occupancy, you will need to comply with the special requirements in Chapter 5 Special Occupancies.

Some examples of special occupancies where you may be installing PV are:

- hazardous locations
- commercial garages
- aircraft hangars
- gas stations (boom!)
- bulk storage plants
- facilities using flammable liquids
- health care facilities
- assembly occupancies (for over 100 people)
- theaters
- amusement parks
- carnivals (fun!)
- motion picture studios
- motion picture projection rooms
- manufactured buildings

- agricultural buildings (what the hay?)
- mobile homes (classy!)
- recreational vehicles
- floating buildings
- marinas (wet!)

Just remember, if you are installing PV in a special place, such as one of the locations listed above, you should take Chapter 5 into consideration.

Chapter 6 Special Equipment

Article 690, which is what most of this book covers, is the most important article in Chapter 6 **Special Equipment**, and there are more articles that may relate to PV installations that we shall mention, such as:

625 Electric Vehicle Power Transfer System (bidirectional charging/discharging)
645 Information Technology Equipment
646 Modular Data Centers
647 Sensitive Electronic Equipment
670 Industrial Machinery
680 Swimming Pools, Fountains, and Similar Installations
682 Natural and Artificially Made Bodies of Water
692 Fuel Cell Systems
694 Wind Electric Systems (our friends)

If your PV system incorporates wind, fuel cells, or electric car charging or is a floating PV system (floatovoltaics), you should look to other articles in Chapter 6 **Special Equipment**.

Chapter 7 Special Conditions

There are many other special conditions besides the interconnection to the grid, which we already covered in Chapter 9 of this book, where we covered Article 705. We also already covered other special conditions in this book in Chapter 10 where we covered energy storage related articles, such as Article 706 Energy Storage Systems and, Article 710 Stand-Alone Systems. Other articles in Chapter 7 Special Conditions that we may have to reference include:

702 Optional Standby Systems (see page 261)
720 Circuits and Equipment Operating at Less than 50V
750 Energy Management Systems (see pages 222–228)

Chapter 8 Communication Systems

Chapter 8 Communication Systems is not subject to the requirements of Chapters 1–7 of the NEC except where referenced in Chapter 8. These systems are separate from PV systems, although they may be powered by solar energy. As solar and energy storage integrate more with the internet, and there is talk of using blockchain technology, AI, and IT for energy and power transactions, we may have to increase the length of this chapter in the future by a few kilobytes.

Chapter 9 Tables

Chapter 9 tables are referenced throughout the NEC. We will look at a few tables that relate to PV system design that will let us know about voltage drop and how many conductors we can physically fit in conduit.

Chapter 9 Table 1 Percent of Cross Section of Conduit and Tubing for Conductors and Cables is used to determine how much of the cross-sectional area of a conduit can be used for wire and how much extra space needs to be left over for air, cooling, and pulling wires. This table is really about geometry and how many circles can fit in a circle. There are three categories in this table. One conductor in a conduit, which is unusual, can take up 53% of the space inside of the conduit. Two conductors in a conduit, which is also not common, can only take up 31% of the cross-sectional area in the conduit. When we have three or more conductors in conduit, which is usually the case, we can take up 40% of the cross-sectional area of the conduit.

Chapter 9 Table 4 Dimensions and Percent Area of Conduit and Tubing is used to determine the interior cross-sectional area of conduit and can be used with Chapter 9 Table 1. This table covers many types of conduit and is many pages long. Additionally, this table gives us the data for the percentages of cross-sectional area that are required by Chapter 9 Table 1, so we do not need to even look at Chapter 9 Table 1 when we are using Chapter 9 Table 4.

For example, the common conduit used for residential PV projects is 3/4-inch EMT and we can use Chapter 9 Table 4 to see that 40% of the interior cross-sectional area of **3/4-inch EMT is 0.213 square inches**. This is how much space we can use without wires and we will see how much space the wires take up next.

Chapter 9 Table 5 Dimensions of Insulated Conductors and Fixture Wires can be used to determine the cross-sectional area dimensions of wires. Like Chapter 9 Table 4, this table is many pages long. This table is also used with Chapter 9 Table 4.

An example of the dimensions of a commonly used wire in the PV industry would be using this table to determine the cross-sectional dimensions of 10 AWG THWN-2. We go down the left column until we see THHN, THWN, THWN-2, which all have the same dimensions, and we can see that **10 AWG THWN-2 has a cross-sectional area of 0.0211 square inches.**

If we use our examples from Chapter 9 Tables 4 and 5, we can divide the cross-sectional area of 3/4-inch EMT, which is 0.213 square inches, by the cross-sectional area of 10 AWG THWN-2, which is 0.0211 square inches, and we get:

0.0213/0.0211 = 10.1 conductors

We have to round down to 10 conductors in this example to fit in this conduit. We will see a little later in this chapter that we can use Informative Annex C to cut down on math.

Rounding up for number of conductors in conduit: **Chapter 9 Note 7** [a.k.a. 9.0(7)] tells us that we can **round up if we are 0.8 over a whole number** when calculating the number of conductors that will physically fit inside a conduit.

People in the field do not like using the maximum number of conductors that can fit in a conduit, because a tight fit can be very difficult to work with. It is recommended to have extra space. This is where there is often a disconnect between the engineers designing the PV system and the swearing installers in the field trying to cram conductors in conduit.

Chapter 9 Table 8 Conductor Properties is used often to calculate voltage drop, but there are also other convenient properties of this table. In our discussion of wire sizing in Chapter 12 of this book, we will use Chapter 9 Table 8 for calculating voltage drop, using the resistance values in ohm per kilofoot (kFT) in order to determine the resistance of a wire.

For instance, we can see that an uncoated copper-stranded 10 AWG wire has a resistance of 1.24 ohms per kFT. This means that 1000 feet of 10 AWG wire has a resistance of 1.24 ohms, so 500 feet would have half of that. One thing in this table that often confuses people is the uncoated vs. coated column in this table. The **coated wire is usually a copper wire that was dipped and has a tinplating on it to help with corrosion.** We most often do not use coated wire and this **coating has nothing to do with the wire having insulation**, as is often confused.

Chapter 9 Table 8 also has the metric cross-sectional area of the wire. **In most of the world, square mm cross-sectional area is used to designate a wire's dimensions** rather than the AWG system. Here we can use this table to convert.

We can also see wire dimensions in circular mils. **A circular mil is the cross-sectional area of a circle with a diameter of 1/1000 of an inch** (very small). Often **larger wires are measured in thousands of circular mils (kcmil).**

Chapter 9 Table 9 Alternating-Current Resistance and Reactance for 600-Volt Cables, 3-Phase, 60Hz, 75°C—Three Single Conductors in Conduit is used to calculate ac voltage drop for larger conductors. With larger conductors, there is a greater difference between ac and dc with regards to voltage drop.

Skin effect is when the alternating current tends to ride on the outer "skin" of the wire with larger wires, thus making ac less efficient than dc with larger wires. This is one reason that larger wires are run in parallel, such as when you see multiples of three wires running along huge utility power poles. This is also why there are high voltage dc (HVDC) transmission lines popping up around the world, with some over 1MV!

There are also more variables with these calculations, such as the type of conduit, the power factor, and copper vs. aluminum wire.

With wires smaller than 250 kcmil, the difference between using Chapter 9 Table 8 and Chapter 9 Table 9 is usually less than 5% more voltage drop.

Table 10 deals with the number of strands in typical cables. The most common conductors use Class B stranding. Standard terminals are designed for Class B stranded conductors. For higher strand counts, as is required for tracking systems in **690.31(C)(4) Flexible Cords and Cables Connected to Tracking PV Arrays** page 115 and **Table 690.31(C)(4) Minimum Wire Strands**, special connectors are required to properly terminate the conductors [see 690.31(C)(5), page 116]. Most PV modules have Class C or similar stranded PV wire for the module leads.

Informative Annexes

Further in the back of the NEC are Informative Annexes and Informative **Annex C Conduit and Tubing Fill Tables for Conductors and Fixture Wires of the Same Size** (long name) is especially useful for determining how many conductors fit in a conduit, in many cases avoiding the math involved with using the tables in Chapter 9.

If we wanted to arrive at fitting ten 10 AWG THWN-2 conductors in a 3/4-inch EMT conduit, we can look for THWN-2 in the left column of Informative Annex C and, a few pages in, we see THWN-2, we match 10 AWG with 3/4-inch EMT, and can see that ten conductors fit. Informative Annex C does not work for conductors of different sizes within the same conduit.

Informative Annexes are not part of the requirements of the Code but contain helpful shortcuts.

Index

Last but not least in the NEC is the Index, which contains everything from ac to zones. Become familiar with the index and use it often, especially if you are planning on taking an open book NEC exam or need to fall asleep fast without damaging your liver. The smartest people know how to use the index.

Since we are talking about the index here, "index" shall be in the index of this book.

We have now covered PV and the NEC. The next and final chapter will use the NEC by working out wire sizing examples.

12 PV Wire Sizing Examples

Wire sizing has been so complicated that many experts disagree on how to correctly size a wire and many books have conflicting methods for how to do it properly. This is also why there are no wire sizing programs on the internet that work for anything besides voltage drop. If someone was good enough at coding to make a good wire sizing program, they would be too busy at Google making $333k per year instead of making NEC wire sizing programs.

In sizing a wire there are many different checks that should be done. Some of the checks seem so obvious that they are usually skipped, while others are sometimes just given a brief statement such as "then check that the overcurrent device satisfies Article 240 Overcurrent Protection."

We will give a few examples of wire sizing and then let you practice on your own. Practice makes perfect. Someone who sizes PV wires every day will often skip most checks, since they know from experience which check will determine the wire size in their particular situation.

It is recommended that you sit down with your NEC and use the following tables to practice these wire sizing exercises. You can also photocopy or screenshot the following tables and rules for your wire sizing kit:

310.16 Ampacities of Conductors NOT in Free Air Based on 30°C
310.17 Ampacities of Conductors in Free Air Based on 30°C
310.15(B)(1)(1) Ambient Temperature Correction Factors Based on 30°C

DOI: 10.4324/9781003189862-13

310.15(C)(1) Adjustment Factors for More than Three Current Carrying Conductors
240.4(D) Small Conductor Rule
240.6(A) Standard OCPD Sizes

Become an expert yourself or hire Bill and Sean to do your wire sizing at attorney-like prices. The difference with Sean and Bill vs. attorneys is that Sean and Bill have souls.

Legend

COU = Conditions of Use (adjustment and correction factors)
Ampacity = Conductor's ability to carry current
Imax = Maximum Circuit Current [690.8(A), page 35]
Icont = required ampacity for continuous current = Imax × 1.25 [690.8(B)(1), page 42]
OCPD = Overcurrent protection device

Wire Sizing Example 1: Inverter Output Circuit Wire Sizing

Given the following information:

- Inverter continuous rated output current = **10A**
  - If it were on a house, then 240V × 10A = 2.4kW inverter

- Number of current-carrying conductors in conduit **2**
  - Do not include ground or balanced neutral

- ASHRAE 2% high temperature from www.solarabcs.org/permitting = **40°C**
- Distance above roof conduit in sunlight = 1 inch (irrelevant info using 2023 NEC)
- Terminal temperature limits = **75°C**
  - Terminals are what we attach ends of wires to.

- Wire type to be used = THWN-2
  - **90°C** rated wire
  - We can see this in Table 310.15(B)(16)
  - -2 at end of wire designation means 90°C rated

Discussion

Defining current for this exercise:

690.8(A)(1)(c) [page 40]

> Maximum circuit current = continuous rated output current = 10A = Imax

690.8(B)(1) [page 42]

> Required ampacity for continuous current = Imax × 1.25 = 12.5A = Icont

This example and these steps use a 90°C rated wire and 75°C rated terminals.

We are going to break this down into 10 steps (at least it's not 12 steps—although you might need to recover with a 12-step program after we are done). Some of the steps will seem useless in most cases, but it is possible to have a 75°C rated wire with 90°C rated terminals, although we have never seen it happen. We could make fewer steps and then find an unusual exception where the four-step process does not work.

10 Steps (Almost as Effective as a 12-Step Program—This Is Ampacity Anonymous)

Step 1 is really without wires (we are powerless, and our electrical systems are unmanageable). If you end up picking a wire that is too small in one of the steps, then you go to a higher size.

The way Sean does it is he puts the following 10 steps in an editable document and then follows the steps as seen below. When he is consulting, he finds that over 90% of people calculate wrong and end up using a more efficient wire than the minimum required, which is probably good in the long run.

(1) Round up Icont to fuse size
(2) Pick conductor size [perhaps from intuition, Tables 310.16, 310.17, or 240.4(D)]
(3) 75°C ampacity
(4) 75°C ampacity ≥ Icont good!
(5) 75°C ampacity ≥ OCPD good!

(6) 90°C ampacity
(7) 90°C ampacity ≥ OCPD good!
(8) 90°C ampacity × COU deratings = COU derated wire
(9) COU derated wire ≥ Imax good!
(10) COU derated wire round up to OCPD ≥ OCPD from step 1 good!

Working the 10 Steps with Our Example 1

(1) Round up Icont to fuse size
- 12.5A rounds up to 15A
- 240.6

(2) Pick conductor size
- Educated guess or from 240.4(D)
- 14 AWG copper is smallest wire from 240.4(D)

(3) 75°C ampacity (75°C terminals)
- 14 AWG = 20A
- Table **310.15(B)(16) conduit** or (B)(17) if free air

(4) 75°C Ampacity ≥ Icont good!
- 20A ≥ 12.5A good!
- If not good, increase the conductor size here

(5) 75°C Ampacity ≥ OCPD good!
- 20A ≥ 15A good!
- If not good, increase the conductor size

(6) 90°C ampacity (90°C rated wire)
- 14 AWG = 25A
- **Table 310.15(B)(16)** conduit or (B)(17) if free air

(7) 90°C ampacity ≥ OCPD good!
- 25A ≥ 15A good
- If not good, increase the conductor size here

(8) 90°C ampacity × COU deratings = COU derated wire
- COU = Conditions of Use
- COU deratings from 310.15(B)(1)(1) and 310.15(C)(1)
- 310.15(B)(1)(1) 40°C for 90°C rated wire = 0.91 derating
- 310.15(C)(1) no derating for two current-carrying conductors
    - Do not count balanced neutral and ground
- 25A wire × 0.91 = 23A **rounded to nearest whole number**

(9) COU derated wire ≥ Imax good!
- 23A ≥ 10A

(10) COU derated wire round up to OCPD ≥ OCPD from step 1 good!
- 23A wire rounds up to 25A
- 25A ≥ 15A

Conclusion: 14 AWG wire satisfies the requirements of the code. In this case, however, most people would use a larger wire due to voltage drop. Ten years ago, everyone used a 10 AWG wire, and today many people still use a 10 AWG wire, but the less expensive PV 12 AWG wire is becoming more common, and the 14 AWG wire does not violate the NEC.

Step 10: Rounding Up Wire to Common Overcurrent Device

One of the most difficult concepts for people to get is that, in step 10, we **can take the ampacity of the wire and round up to the next standard overcurrent protection device size. This is not saying that the conductor has a higher ampacity**; this is just saying that a wire derated to 23A can be protected by a slightly larger sized OCPD and still be safe. This rounding up comes from 240.4(B) Overcurrent Devices Rated 800A or less (also seen on page 53). For overcurrent devices over 800A, you do not get to round up the ampacity of the conductor over the overcurrent device size, which seems to make more sense.

Wire sizing for voltage drop is a good idea, but it is never a Code issue for PV with the NEC. We will do voltage drop calculations later in this chapter, after we focus on Code-compliant wire sizing.

Wire Sizing Example 2: PV String Circuit Wire Sizing (a 2023 NEC PV String Circuit Is a Type of PV Source Circuit)

Sizing a PV string circuit given the following information:

- Isc = 8A
- Number of PV string circuits in a conduit = 20
- ASHRAE 2% high temperature from www.solarabcs.org/permitting = 40°C

- Distance above roof conduit in sunlight = 1 inch
- Terminal temperature limits = 75°C
- Wire type to be used = THWN-2

Discussion

Defining current:

690.8(A)(1)(a)(1) [Page 35]
Maximum circuit current = Isc × 1.25 = 8A × 1.25 = 10A = Imax (Imax is different from, and not to be confused with, Imp)

Required ampacity for continuous current:

690.8(B)(1) [page 42]
Required ampacity for continuous current = Imax × 1.25 = 12.5A = Icont
(Icont = Isc × 1.25 × 1.25 = Isc × 1.56)

Conductors:

THWN-2 = 90°C rated wire and we are using 75°C terminals as mentioned.
20 PV source circuits = 40 current-carrying conductors

Working the 10 Steps with Our Example 2

(1) Round up Icont to fuse size
  - Icont = 12.5A rounds up to 15A fuse as per 240.6
(2) Pick conductor size
  - 15A fuse requires at least 14 AWG copper as per 240.4(D)
(3) 75°C ampacity (75°C terminals)
  - 75°C 14 AWG = 20A as per Table 310.16
(4) 75°C Ampacity ≥ Icont good!
  - 20A ≥ 12.5A good!
(5) 75°C Ampacity ≥ OCPD good!
  - 20A ≥ 15A good!
(6) 90°C ampacity (90°C rated wire)
  - 14 AWG = 25A per Table 310.16

(7) 90°C ampacity ≥ OCPD good!
   - 25A ≥ 15A good

(8) 90°C ampacity × COU deratings = COU derated wire
   - 310.15(B)(1)(1) 40°C for 90°C rated wire = 0.91 derating
   - 310.15(C)(1) for 40 conductors in conduit = 40% = 0.4
   - 25A wire × 0.91 × 0.4 = 9A rounded to nearest whole number

(9) COU derated wire ≥ Imax good!
   - 9A is *not* ≥ 10A, so go back to use next larger wire 12 AWG
   - **12 AWG = 30A** per Table 310.16
   - **30A** × 0.91 × 0.4 = 11A rounded to nearest whole number
   - **11A ≥ 10A** (notice we are not using Icont here)

(10) COU derated wire round up to OCPD ≥ OCPD from step 1 good!
   - **11A wire rounds up to 15A** as per 240.6
   - 15A ≥ 15A

Conclusion: **12 AWG satisfies the requirements of the Code here**. It is interesting to note that the condition of use rated wire is 11A and we can round that up to 15A and have an **11A wire protected by a 15A overcurrent protection device!** If you go to Europe, you will see that their wires can carry more current for the same size wire than AWG wires can. We have a buffer of protection built into our wires that will let us seem to deny common sense and round up a wire's ability to carry current.

Would we use a 12 AWG wire here in reality? I think I would use a 10 AWG wire, just to be safe and simple. We do not want to push our luck here with what we have learned in Chapter 12.

Voltage Drop (Rise)

When it comes down to voltage drop (rise), what we really want to know is how much money our wire will save for us if we invest more money in the wire. There may be complex calculations, which would have to include tilt, azimuth, copper prices, aluminum prices, PV prices, soiling, labor prices, inverter prices, incentive type, PV to inverter ratio, and weather. To perform those calculations, it is recommended to use complex software and perhaps to hire a team of engineers (or do what everyone else does and use a 10 AWG wire).

Voltage Rise

The reason that we keep mentioning voltage rise in this book, is because the voltage must be at least slightly higher where the electricity is coming from, than where it is going to. The difference in voltage is determined by the resistance of the conductor and the current (V=IR). With a source, the voltage will be higher at the source, so with your typical interactive inverter output circuit, the voltage will be higher at the inverter than at the point of interconnection, and will rise even more as the sun gets brighter (current goes up). This is really voltage drop, but it appears as voltage rise from the inverter's point of view as the inverter pushes current towards the grid. According to an interactive inverter, the grid is a load!

For the purposes of this book, we will use the maximum output current of the inverter, which is being very conservative, since most if not all the energy generated from a PV system is going to be less than the maximum output current. For PV source circuits, we will use current at maximum power (Imp), which is considerably less than the currents we used to calculate Code-compliant wire sizes and is more than we will often see on a PV source circuit.

Some designers will use 80% of these numbers as a rule of thumb, since most of our energy is made when it is not a cold, windy, bright summer noon (optimal PV conditions). We will use Imp and inverter maximum output current for this book, which is conservative and leads to less energy loss over the year than voltage drop percentage in the calculation.

If you are performing voltage drop calculations for a job that you have won a bid on or are bidding on, you should carefully read the requirements of the request for proposal.

We will use a simple calculation to arrive at an AWG wire size given the following information:

Voltage = 240V
Current = 16A
Voltage Drop Percentage = 2%
Distance from inverter to interconnection = 200 feet

Here is the formula that can be used with Chapter 9 Table 8 of the NEC

$$\text{Ohms/kFT} = (5 \times \% \times V)/(I \times L)$$

Ohms/kFT will give us an AWG wire size in Chapter 9 Table 8
5 is a constant derived from (1000FT/kFt)/100%/2 wires in a circuit)
% is the percentage, so we use 2 (not 0.02) for 2%
V is the operating voltage, which is 240V at your house
I is the current of the inverter in this case, which is 16A for a 3.8kW inverter
L is the 1-way distance in feet which is 200 FT

We will plug it in to the equation:

$$\text{Ohms/kFT} = (5 \times \% \times V)/(I \times L)$$

$$\text{Ohms / kFT} = (5 \times 2\% \times 240\text{V})/(16\text{A} \times 200\text{FT}) = 2400/3200 = 0.75 \text{ ohms / kFT}$$

If we **look up 0.75 ohms/kFT in Chapter 9 Table 8** we see that an uncoated 6 AWG copper wire will have a resistance of 0.491 ohms/kFT and a smaller 8 AWG stranded copper wire will have a resistance of 0.778 ohms/kFT.

Since voltage drop is not a Code issue here, you can choose to round up or down from a 6 AWG or an 8 AWG wire.

This calculation will work for ac and dc wires because the values in Table 9 are essentially the same for ac circuits running at unity power factor. If you are using a large wire for ac and running the circuits at a power factor of 0.85 (may be required occasionally by utilities for grid support), then the values in Table 9 differ from those in Table 8. It's best to get an engineer involved for larger systems as these calculations can get complicated.

To use these calculations for 3-phase power, just remember that there is a benefit to using 3-phase that is proportional to the square root of 3 (about 1.73). If we divide the square root of 3 by 2 we get 0.886, so we will have 88.6% of the resistance with 3-phase wires or we can multiply our ohms/kFT answer by 0.866. In the example we

Figure 12.1 Nikola Tesla demonstrates how to truly understand 3-phase in 1899.

used, instead of 0.778 ohms/kFT, we could use a wire that is 0.778 × 0.866 = 0.67 ohms per kFT for 240V 3-phase.

The reason we divide the square root of 3 by 2 is because, with 3-phase, our currents are not directly opposing each other (square root of 3) and we are converting from a calculation that is from single phase power where we have to double the one-way distance of our wire to calculate the resistance of a circuit.

A circuit is a circle and if you are going to have your inverter 200 feet from the interconnection, you need to run electrons through 400 feet of wire and will have 400 feet = 0.4 kFT of resistance. With 3-phase, you will need to have current on three wires, but it will be less current, since the currents are 120 degrees out of phase with each other.

Some people say that 3-phase power takes more than a lifetime to truly understand, but if Tesla (a crazy genius) could figure out how 3-phase power worked all on his own, you can too!

Thank you for reading this book! Sean and Bill.

Index

Note: Page numbers in *italic* refers to Tables.